Lecture Notes in Economics and Mathematical Systems 598

Aljaž Ule

Partner Choice and Cooperation in Networks

Theory and Experimental Evidence

Springer

Dr. Aljaž Ule
Roetersstraat 11
1018 WB Amsterdam
The Netherlands
a.ule@uva.nl

ISBN 978-3-540-73015-6 e-ISBN 978-3-540-73016-3

DOI 10.1007/978-3-540-73016-3

Lecture Notes in Economics and Mathematical Systems ISSN 0075-8442

Library of Congress Control Number: 2007938262

Production: LE-TEX Jelonek, Schmidt & Vöckler GbR, Leipzig
Cover design: WMX Design GmbH, Heidelberg

Printed on acid-free paper

9 8 7 6 5 4 3 2 1

springer.com

Starim staršem

Preface

In this book the methods of game theory and experimental economics are applied to investigate the dynamics of cooperation in endogenously formed networks. This investigation is motivated by two observations regarding a traditional assumption in game theoretic models of social dilemmas that participants in such a dilemma have no other option but to engage. One observation is that these models fail to capture an important element of human interaction, as people are in general free to select their interaction partners. The other observation is that partner choice may increase cooperation through exclusion of free-riders. Intuitively, people seek helpful other people for interaction. Cooperative individuals may therefore acquire more interaction partners and earn higher benefits from their interaction network than the individuals who try to free-ride on the others' cooperative contributions. Consequently, cooperative behavior may become profitable, inducing even the selfish individuals to cooperate.

Thus motivated, a social dilemma with partner choice is introduced in the book. It allows exploration of simultaneous dynamics of the network structure and cooperative behavior on this structure. Its analysis shows that partner choice leads to networks where free-riders are likely to be excluded and confirms our intuition that free partner choice strongly facilitates cooperation.

The research presented in this book has been conducted while I was a graduate student at the Tinbergen Institute and at CREED, the Center for Research in Experimental Economics and Political Decision-Making at the University of Amsterdam. It benefited greatly from the academic guidance by my supervisor Arno Riedl, credited among else with co-authoring the research in Chap. 5, and from the assistance by my promotor Arthur Schram. Much support has been provided also by my colleagues in Amsterdam, Barcelona and Ljubljana and by my parents, grandparents and my brother.

Amsterdam, December 2007 *Aljaž Ule*

Contents

1

Motivation and Outline

In many social and economic situations groups of agents, such as people or firms, face a conflict between their individual self interests and the common interests of the group as a whole. Such situations are referred to as the *social dilemmas* and have long been a theme of research in social sciences (see Dawes [1]). Agents are said to *cooperate* in social dilemmas when their decisions support a common good of their reference group. If instead they pursue their individual egoistic interests, in conflict with the common good, then they are said to *free-ride*, or *defect*.

There are many examples of social dilemmas that economists find relevant, such as arms races or trade negotiations between countries, pollution within cities, knowledge and innovation sharing between private firms, provision of freely available scientific knowledge, and use of public resources such as forests and fisheries [2, 1, 3, 4]. Failure to cooperate is usually associated with a loss of efficiency and the economic profession is interested in identifying cases of successful cooperation, and the mechanisms through which cooperation is achieved.

As an example, consider the following social dilemma that often faces a group of graduate students. On Sunday evenings one of the students invites a limited number of others for a dinner at her or his house. The guests are asked to contribute a bottle of a nice wine, while all else is provided by the host. It is always an enjoyable event, and rarely does anyone decline an invitation. Hosting a dinner, however, involves substantial costs in terms of money, effort and time, as well as the risk that the culinary effort would fail. Many, hence, shy away from hosting while maintaining their presence as guests. In other words, all students benefit if everyone occasionally hosts a diner, but if each student prefers to free-ride on the cooperative culinary efforts of other students then there are no dinners. Hence the dilemma. Interestingly, the student dinners often take place, but the hosts typically select their guests only among the students that have hosted a dinner in the recent past.

Standard theory typically predicts no cooperation. The free-riding hypothesis asserts that agents always act in compliance with their individual self interests and therefore fail to achieve any cooperation in social dilemmas. Thomas Hobbes, for example, argued already in the 17th century that cooperation must be externally enforced to avoid societies plunging into universal free-riding and anarchy [5]. The free-riding hypothesis is, however, not consistently supported by empirical evidence. Anthropological literature has documented a number of cases of cooperation in societies that lack formal institutions of enforcement [6, 7]. People seem to be especially successful in overcoming social dilemmas in small groups and in networks of reciprocal exchange [8, 9, 10].

Casual observation is sufficient to realize that people often cooperate. We all encounter social dilemmas in our daily social interactions with other people. Information sharing, organization of social events, and exchange of financial and other help are common instances of social dilemmas that take place in networks of social relations, referred to as the *social networks.* If all people only pursued their immediate private interests we would have little reason to establish friendly relations. Yet, most people do keep friends and enter willingly with them into interactions with the character of a social dilemma. In these interactions many behave cooperatively, sharing advice and valuable information, spending time to give help, and exchanging financial assistance, all this in absence of an external enforcing authority. Such is the apparent association of social networks with cooperation that some social scientists consider them as a sort of capital of individuals and groups, that is, their "social" capital [11, 12]. Given this evidence it is in order to ask: *what drives cooperation in social networks?*

Several answers have been proposed in the theoretical literature. One line of research investigates the role of reputation. Standard models of economic action assume that atomized actors interact in anonymous markets. Cooperation may indeed be difficult to achieve in such an anonymous setting. In social networks, however, interactions are repeated and non-anonymous. It has been analytically shown, for example, that cooperation can be achieved within pairs of rational and selfish agents if they repeatedly interact with each other [13, 14]. Another line of research, suggests that cooperation may be further enhanced if interaction is structured [15, 16]. This conjecture is analytically investigated by Nowak and May [17] and Eshel et al. [18] who compare structured groups, where agents interact through networks, with unstructured groups in which each agent interacts with all other agents. Assuming that behavior spreads by means of imitation they show that cooperation may survive in structured groups but not in the unstructured ones. However, the results of laboratory experiments do not support all these claims. Experimental studies provide evidence for the result that pairs of players achieve high cooperation in repeated interaction, but suggest that it cannot be generalized to groups with three or more agents [19, 20]. Further experimental evidence suggest that cooperation is equally hard to achieve in structured and in unstructured groups [21, 22].

A more promising result for explaining cooperation in networks comes from a third line of research that investigates two-agent social dilemma situations with voluntary entry. In such a situation each of the agents decides whether or not to enter with the other into a relationship in which they would face a social dilemma. The interaction takes place only if both players decide to enter, while otherwise both players take an outside option. Orbell et al. [23] argue that cooperators overestimate the propensity of others to cooperate and, compared to free-riders, relatively more frequently participate in the social dilemma. They conjecture that such self-selection leads to high cooperation in established bilateral relations. Experimental evidence supports this conjecture, both in the two-agent and in the multi-agent setting [24, 25].

In this book we take the investigation of cooperation in networks one step further in a natural way. We consider *endogenous networks*, that emerge as a result of agents' endogenous choices of the partners for interaction, and investigate the simultaneous evolution of cooperation and of network structures in social dilemmas. Our main motivation to study cooperation in endogenous networks is that many real-life situations are better captured by the assumption that the networks are established by the agents themselves, instead of being imposed on the agents exogenously. It seems intuitive that people tend to manipulate social relationships on the basis of observed past behavior. The endogenous choice of interaction partners may, however, be explanatory for the observed cooperation in social networks. For example, cooperators may form social ties with other cooperators and exclude those that cheat. In repeated situations cooperators can identify each other and establish highly cooperative network clusters. As the sizes of cooperative clusters increase the tolerance of free-riders may decline. As a consequence of the threat of exclusion, the free-riding agents may switch to cooperation.

Our analysis relies on the tools of game theory, which is suitable to study both cooperation and network formation. We use a non-cooperative game to model a social dilemma on an endogenous interaction network. We investigate the game using three different approaches. We begin with a standard equilibrium analysis of the game. We then set up laboratory experiments to study actual behavior in this game. Finally, we use computer simulations to investigate the dynamics of adaptive behavior under the assumption that the game is played by boundedly rational agents.

Each of our three approaches is motivated, presented and discussed in a separate chapter. In Chap. 4 we focus on equilibrium analysis, in Chap. 5 on laboratory experiments, and in Chap. 6 on computer simulations. Each of these chapters is structured as follows. In the introduction to the chapter we motivate the methodology and discuss our research in view of the related literature. We then describe the details of the approach and specify our results. We proceed with a critical discussion of our assumptions and wrap up the chapter with a summary of the main conclusions. The mathematical proofs that are not essential to grasp the arguments in the main body of the book are collected in Appendix A.

The three lines of literature related to our three approaches are quite distinct and instead of giving a combined overview we discuss each line separately in the appropriate chapter. Conclusions are also discussed in the corresponding chapters and we do not restate them in a separate summary chapter. Therefore, each of the Chaps. 4–6 can be read in conjunction with the definition of our model in Chap. 3 and independently of the rest of the book. The three chapters are, however, fairly interrelated and are organized in a logical sequence.

We give next a brief overview of the different chapters and their main conclusions. In Chap. 2 we describe the basic concepts of the theory of non-cooperative games, the mathematical theory of networks, and the theory of Markov chains. In Chap. 3 we introduce the model of a social dilemma on endogenous networks that we study throughout the book: the *network dilemma game*. The network dilemma game is, essentially, a multi-player Prisoner's dilemma game in which the interaction network depends on the player's choices of their interaction partners. We conclude Chap. 3 with a discussion of our model.

In Chap. 4 we study subgame-perfect equilibria of various finitely repeated network dilemma games. We focus on games in which players are constrained in the number of other players with whom they can interact. Our main conclusion is that for many such games a subgame-perfect equilibrium exists in which all players cooperate during all but the final few periods, given that the number of periods is sufficiently large. A sufficient condition for existence of such a subgame-perfect equilibrium is, roughly, that the set of Nash equilibrium networks includes several networks with no isolated players, but not the complete network. These results imply that it is possible to achieve cooperation in a subgame-perfect equilibrium of a finitely repeated Prisoner's dilemma game if it is played on an endogenous network.

In Chap. 5 we describe an experimental investigation of behavior in finitely repeated network dilemma games. Our analysis in Chap. 4 predicts no cooperation in games with no interaction constraints. In contrast, we observe in our related experimental games, on average, high and increasing cooperation during all but the last few periods. Our main conclusions are that people are willing to enter into risky social dilemma situations, that they use costly exclusion as a means of retaliation for free-riding, and that partner choice is a powerful means for promoting cooperation. Our further analysis suggests that behavior in our experimental games cannot be explained by altruism or inequity aversion of our subjects.

In Chap. 6 we investigate the dynamics of cooperation and partner choice of free-riders in groups of fictitious players with "limited forward looking" who repeatedly play the network dilemma game with no interaction constraints. We assume that players choose simple behavioral rules which maximize their payoffs across a limited number of future periods. In this sense the players are non-myopic best-responders. Our main conclusion is that limited forward looking is sufficient to achieve modest amounts of cooperation in some network dilemma games. We also observe that cooperation is further facilitated if players are highly altruistic, but only if they are forward looking.

Finally, in this book we want to advance a few general messages, in addition to our specific results in Chaps. 4–6. First, in the traditional models of social dilemmas in economics the agents have no other option but to participate in a dilemma. This may be a fair description of the situation that inspired the use of the term "Prisoner's dilemma", but there are also many real-life situations in which social or economic agents have more freedom to select with whom they are willing to interact. The results of our Chaps. 4 and 5 suggest that economic models of such situations are not complete unless they include the endogenous choice of interaction partners.

Second, in Chap. 6 we study a model of agents with limited foresight. People do not have an unlimited ability of predicting behavior of others, and assuming this can lead to empirically inaccurate predictions about their behavior in dynamic situations. On the other hand, it also seems unreasonable to think that people have no ability to predict the consequences of their actions. We therefore suggest that realism of models of human decision-making would improve if humans were modeled as agents with limited foresight.

All in all, our goal in this book is to take our understanding of how individuals behave in social dilemma-type situations one step further. Our results suggest that cooperation is much more likely among individuals who are free to decide with whom they enter in a social dilemma-type situation. The methods that we develop in this book may facilitate analysis of future models of decision-making on endogenous networks.

2

Basic Concepts of the Theories of Games, Networks and Markov Chains

This chapter introduces the formal concepts that we use in the rest of the book. Section 2.2 describes the elements of the non-cooperative game theory. Section 2.3 describes the elements of the mathematical theory of networks. Section 2.4 describes the basic concepts and results of the theory of Markov chains.

2.1 Notation

The following notation is used throughout the book. For any set S and any element $s \in S$ let $S\backslash\{s\}$ be the set obtained by removing s from S. For any set S and any subset $R \subseteq S$ let $S\backslash R$ be the set obtained by removing from S all elements in R. For any two sets S_1 and S_2 let $S_1 \times S_2 = \{(s_1, s_2) \mid s_1 \in S_1, s_2 \in S_2\}$ be their Cartesian product. Let $N = \{1, ..., n\}$. For any sets $S_1, ..., S_n$ let

$$\times_{i \in N} S_i = S_1 \times ... \times S_n = \{(s_1, ..., s_n) \mid s_1 \in S_1, ..., s_n \in S_n\} .$$

For any positive integer t and any set S let

$$\times^t S = \underbrace{S \times ... \times S}_{t} = \{(s_1, ..., s_t) \mid s_1, ..., s_t \in S\} .$$

Finally, let $\{g^i\}_{i \in N} = \{g^1, ..., g^n\}$, $(p_i)_{i \in N} = (p_1, ..., p_n)$, $(p_{ij})_{j \in N} = (p_{i1}, ..., p_{in})$ and $(t^\eta)_{\eta=0}^m = (t^0, ..., t^m)$.

2.2 Elements of Game Theory

A non-cooperative game is a description of an interaction between a set of agents, called players. The most basic class of non-cooperative games is that of strategic games. A strategic game describes a situation in which each of the agents chooses his action or a plan of actions once and for all, and these choices are made simultaneously. Each possible profile of actions, one for each player, describes one possible outcome of such game. A solution is a systematic selection criterion over the outcomes.

The definition of a strategic game describes the players, the actions that players can take, and the players' preferences over the possible outcomes of the game.

Definition 2.1. *A finite* ***strategic game*** *is a triple* $\langle N, (A_i)_{i\in N}, (\succsim_i)_{i\in N}\rangle$*, where*

- $N = \{1, 2, ..., n\}$ *is the finite set of* ***players****,*
- *for every player* i*,* A_i *is the finite set of* ***actions*** *available to* i*,*
- *for every player* i*,* $\succsim_i$ *is a* ***preference relation*** *on* $A = \times_{i\in N} A_i$*, reflecting her preferences over* ***outcomes*** A*.*

In a strategic game each player i chooses her action a_i simultaneously with all other players. The outcome of the game is given by the profile of all chosen actions $a = (a_1, ..., a_n) \in A$. The set A is called the set of action profiles and can be seen as the set of the outcomes of the game.

Players

$$N\backslash\{i\} = \{1, ..., i-1, i+1, ..., n\}$$

are called *opponents* of player i.[1] For each action profile a let

$$a_{-i} = (a_1, ..., a_{i-1}, a_{i+1}, ..., a_n)$$

be the profile of actions taken by i's opponents. Let $A_{-i} = \times_{j\in N\backslash\{i\}} A_j$ be the set of all such action profiles. The profile of actions such that player i chooses a_i and her opponents choose a_{-i} is denoted by (a_i, a_{-i}).

Often the preference relation $\succsim_i$ of player i can be represented by a *payoff function* $\pi_i : A \to \mathbb{R}$, in the sense that

$$\pi_i(a) \geq \pi_i(b) \Longleftrightarrow a \succsim_i b$$

for all pairs of outcomes $a, b \in A$.[2] The payoffs are meant to represent player's psychological utilities for various outcomes. Therefore, the payoff function is also called utility function. Two outcomes $a, b \in A$ are *payoff-equivalent* if $\pi_i(a) = \pi_i(b)$ for each i.

---

[1] This may be an unjust term as it gives an impression that players are necessarily adversaries. In a game the player's interests may be conflicting but need not be.

[2] Von Neumann and Morgenstern [26] have shown that this is possible under remarkably weak assumptions about consistency and regularity of preferences. For a comprehensive discussion see also [27].

		Player j C	D
Player i	C	5, 5	1, 7
	D	7, 1	3, 3

Fig. 2.1. An example of a payoff matrix representing a two-player strategic game. This is an instance of a prisoner's dilemma game

Throughout the book we consider non-cooperative finite games with preferences represented by payoff functions. For simplicity we will refer to a finite strategic game with set of players N, action sets A_i, and payoff functions π_i simply as *game* $\langle N, A, \pi \rangle$, where $\pi = (\pi_1, ..., \pi_n)$.

A game with two players can be described conveniently using a table, called *payoff matrix*, like the one illustrated in Fig. 2.1. Player 1 is identified with the rows and player two with the columns. The column to the left of the table lists the actions available to player 1. The row at the top of the table lists actions available to player 2. Each cell of the table is associated with an action of player 1 and an action of player 2 and describes the payoffs that the players earn when each plays the corresponding action. The first number in a cell is the payoff to player 1 and the second number the payoff to player 2.

In the game described by the payoff matrix in Fig. 2.1 each of the players has two possible actions, C and D. If, for example, player 1 chooses C and player 2 chooses D, the payoff to player 1 is 1 and the payoff to player 2 is 7. This game is an instance of prisoner's dilemma games described below.

A standard assumption in game theory is that players are *rational* in the sense that they always select an action that maximizes their psychological utility given their beliefs about the actions being taken by the other players. Rationality is embedded in the notion of a best response mapping. For each player i her best response mapping $B_i : A_{-i} \longmapsto A_i$ is defined by

$$\begin{aligned} B_i(a_{-i}) &= \{a_i^* \in A_i \mid \pi_i(a_i^*, a_{-i}) \geq \pi_i(a_i, a_{-i}) \text{ for all } a_i \in A_i\} \\ &= \arg\max_{a_i \in A_i} \pi_i(a_i, a_{-i}) \ . \end{aligned}$$

For each profile $a_{-i} \in A_{-i}$ the set $B_i(a_{-i})$ consists of those actions of player i which maximize her payoff given that her opponents play a_{-i}. An action $a_i^* \in B_i(a_{-i})$ is a *best response* of player i to actions a_{-i} of her opponents. For notational convenience we define also $B_i(a) \equiv B_i(a_{-i})$ for $a = (a_i, a_{-i})$. If $a_i \in B_i(a)$ we say that player i is playing her best response in a.

The commonly used solution concept for strategic games is that of the Nash equilibrium [28]. It characterizes action profiles which may be seen as steady states of the game. Each rational player i who knows that her opponents are playing a_{-i} is willing to play a_i if and only if it is a best response to a_{-i}. In this sense rational players are willing to conform to an outcome $a \in A$ if each of them is playing her best response in a and each knows that a will be chosen.

Definition 2.2. *A* ***Nash equilibrium*** *of a strategic game $\langle N, A, \pi \rangle$ is an action profile $a^* \in A$ such that*

$$a_i^* \in B_i^*(a) \textit{ for every player } i \ .$$

We may restate the definition as follows: $a^* \in A$ is a Nash equilibrium if and only if for each i,

$$\pi_i(a_i^*, a_{-i}^*) \geq \pi_i(a_i', a_{-i}^*) \text{ for all } a_i' \in A_i \ .$$

Profile $a^* \in A$ is a *strict* Nash equilibrium if and only if for each i,

$$\pi_i(a_i^*, a_{-i}^*) > \pi_i(a_i', a_{-i}^*) \text{ for all } a_i' \in A_i, \ a_i' \neq a_i^* \ .$$

2.2.1 Finitely Repeated Games

A non-cooperative game in which not all decisions are made simultaneously can be described as an extensive game. In this book we restrict our attention to extensive games obtained when a strategic game is repeated finitely many times. Such games are called finitely repeated games. We assume perfect recall, that is, at each period of the repeated game, each player is perfectly informed about all actions taken by all players in all previous periods. In game-theoretic terms, we assume complete information.

In each period of the repeated game all players simultaneously choose an action for the constituent strategic game. In each period they may make their decisions using the information about all past choices. Rational players take into account the effect of their current actions on the future behavior of the other players. Repeated games may thus capture phenomena like reputation, threats, and revenge.

Let $\Delta = \langle N, A, \pi \rangle$ be a strategic game. Let $T < \infty$ be the number of repetitions of the game Δ. We refer to the constituent strategic game Δ as the *stage game.*

For each period $t \in \{1, 2, ..., T\}$ the profile of actions at t is denoted by $a^t \in A$. The action of player i in period t is $a_i^t \in A_i$. A t-period *history* is a vector of action profiles in periods $1, 2, .., t$,

$$h^t = (a^1, a^2, ..., a^t) \ .$$

The space of all possible t-period histories is $H^t = \times^t A$. Take t and τ such that $t + \tau \leq T$. For any t-period history $h^t = (a^1, ..., a^t)$ and any τ-period history $\overline{h}^\tau = (\overline{a}^1, ..., \overline{a}^\tau)$, let

$$h^t \oplus \overline{h}^\tau = (a^1, ..., a^t, \overline{a}^1, ..., \overline{a}^\tau)$$

be the $(t+\tau)$-period history given by actions h^t during the first t periods and $\overline{h}^\tau$ during the following τ periods.

The sequence of outcomes in each repetition of the stage game constitutes the outcome path of the repeated game. Let a^t be the period t outcome. The *outcome path* can be represented by the T-period history $h^T = (a^1, ..., a^T)$. We assume throughout the book that players do not discount the future. The aim of a rational player i is to maximize her *total payoff*, given by

$$\Pi_i(h^T) = \sum_{t=1}^{T} \pi_i(a^t) \ .$$

Definition 2.3. *A* ***finitely repeated game*** *is a tuple* $\Delta^T = \langle N, A, \pi, T \rangle$*, representing a* T*-fold repetition of the stage game* $\Delta = \langle N, A, \pi \rangle$.

The behavior of a player in the repeated game is characterized by a strategy, which is a plan of actions that specifies, at the beginning of each period, how her action will depend on the most recent history of play. The behavior of players in repeated games is thus modeled by functions that map histories into actions.

A *strategy* for player i in the game Δ^T is a function σ_i which selects, for any history of play, an element of A_i. Formally,

$$\sigma_i = (\sigma_i^1, \sigma_i^2, ..., \sigma_i^T) \ ,$$

where $\sigma_i^1 \in A_i$ and for $t > 1$,

$$\sigma_i^t : H^{t-1} \rightarrow A_i \ .$$

Let $\sigma^t = (\sigma_1^t, ..., \sigma_n^t)$. A profile of strategies $\sigma = (\sigma_1, \sigma_2, ..., \sigma_n)$ inductively defines an outcome path of the game Δ^T as follows: $a^1(\sigma) = \sigma^1$ and for $t \geq 1$, $a^{t+1}(\sigma) = \sigma^{t+1}(h^t(\sigma))$ where

$$h^t(\sigma) = (a^1(\sigma), ..., a^t(\sigma)) \ .$$

Let S_i be the space of strategies of player i and let $S = \times_{i \in N} S_i$. The repeated game $\Delta^T = \langle N, A, \pi, T \rangle$ can be associated with the strategic game $\langle N, S, \Pi \rangle$, called the *normal form* of the repeated game Δ. The normal form describes the game in which each player chooses a strategy $\sigma_i \in S_i$ and earns the payoff, determined by the corresponding outcome path $h^T(\sigma)$,

$$\Pi_i(\sigma) = \Pi_i(h^T(\sigma)) = \sum_{t=1}^{T} \pi_i(a^t(\sigma)) \ .$$

There is a crucial difference between the repeated game and its normal form. In the repeated game the players may modify their strategy in each period. In the normal form, however, the strategy is chosen once and for all at the outset of the game. The normal form is thus not the perfect representation of the repeated game. However, we can relate its solutions, such as its Nash equilibria, to the solutions for the repeated game.

In short, a strategy profile $\sigma \in S$ is a Nash equilibrium of the repeated game if and only if it is a Nash equilibrium of the corresponding normal form. Formally, let (σ_i', σ_{-i}) denote the strategy profile σ but with the ith component replaced by σ_i'. A *Nash equilibrium* of game Δ^T is a profile σ^* such that for each player i, and any strategy σ_i for i,

$$\Pi_i(\sigma^*) \geq \Pi_i(\sigma_i', \sigma_{-i}^*) \ .$$

The outcome path $h^T(\sigma^*)$ is an *equilibrium path* if σ^* is a Nash equilibrium. We need to distinguish between the Nash equilibria of the repeated game and those of the stage game. We refer to the latter as the *static* equilibria. If an outcome in some period of the repeated game constitutes a static equilibrium, we say it is an *equilibrium outcome.*

2.2.2 Subgame Perfect Equilibrium

The concept of Nash equilibrium for repeated games treats the strategies as choices that are made once and for all at the outset of the first period. As such, it ignores the sequential structure of the repeated game and the fact that players can change their strategies between periods. The profile of strategies is a Nash equilibrium of the repeated game if each player's strategy is optimal against the strategies of the other players along the equilibrium path. Off this path, that is, in case of a deviation, strategies do not need to be optimal. In particular, there is no restriction that in case of a deviation the players should be willing to continue to use their strategies.

We say that a Nash equilibrium of a repeated game is supported by non-credible threats if, in case of some deviation, some rational player does not want to follow her strategy.[3] The equilibrium is supported by credible threats only if, after any history of play, it is in the self-interest of each player to continue using her original strategy. In a group of rational players the equilibrium path is sustainable if each player's strategy is the best response not only along the equilibrium path but also in case of any deviation.

This is the spirit of the notion of subgame perfection introduced by Selten [31] to eliminate the Nash equilibria of finite extensive form games which are sustained only as a result of non-credible threats off the equilibrium path. Selten suggested the solution concept of subgame perfection, which requires a strategy profile to be a Nash equilibrium along the equilibrium path, as well as along all other paths, that is, after any possible history of play.

Consider a finitely repeated game $\Delta^T = \langle N, A, \pi, T \rangle$. Let h^t be a t-period history after some period $t < T$ and let $\Delta^{T-t} = \langle N, A, \pi, T-t \rangle$ be the game which consists of the remaining $T-t$ periods of play. If σ_i is a strategy for i in Δ^{T-t}, let $\sigma_i\,|_{h^t}$ be its continuation for periods $t+1, ..., T$, given the history h^t. Strategy $\sigma_i\,|_{h^t}$ may then be seen as a strategy for player i in game Δ^{T-t}, defined iteratively by: (i) $\sigma_i\,|_{h^t}{}^1 = \sigma_i^{t+1}(h^t)$, and (ii) for any $m < T-t$ and any m-period history $\overline{h}^m$,

$$\sigma_i\,|_{h^t}{}^{m+1}\,(\overline{h}^m) = \sigma_i^{t+m+1}(h^t \oplus \overline{h}^m) \ .$$

The corresponding strategy profile in game Δ^{T-t} is defined by $\sigma\,|_{h^t} = (\sigma_1\,|_{h^t}, ..., \sigma_n\,|_{h^t})$.

A *subgame* of a repeated game Δ^T is any pair (h^t, Δ^{T-t}) with $t < T$. The set of all subgames of a repeated game is generated by combining any possible history with the remaining periods of play.

Definition 2.4. *The strategy profile σ is a* ***subgame perfect equilibrium*** *of the game Δ^T if (i) it is a Nash equilibrium of Δ^T, and (ii) for all subgames (h^t, Δ^{T-t}), $\sigma\,|_{h^t}$ is a Nash equilibrium of Δ^{T-t}.*

Take any, possibly non-equilibrium, strategy profile $\sigma \in S$. Let h^{t-1} be the history in period t. If player i follows her strategy in period t she selects the action $\sigma_i^t(h^{t-1})$. We say that player i *deviated* from σ in period t if her action at t differs from $\sigma_i^t(h^{t-1})$. We say that a deviation from σ occurred at t if the outcome at t differs from the actions selected by σ given the history h^{t-1}.

A *threat* is an outcome path $h^{T-t}(\sigma\,|_{h^t})$ for the subgame (h^t, Δ^{T-t}) selected by σ if a deviation occurs at t. The threat selected by σ in

[3] See [29], p. 4, and [30], p. 93, for some simple examples of Nash equilibria sustained by non-credible threats. See also example 4.3 in Chap. 4.

response to a deviation at t is *credible* if $\sigma|_{h^t}$ is a subgame perfect equilibrium of (h^t, Δ^{T-t}).

A threat keeps players from deviating if and only if, at the end of the game, the deviating players earns no more than what they would have earned had there been no deviation, assuming no other deviations occur. Suppose that player i deviates at t, and that no more deviations occur at t or in any later period. The total payoff of i is then $\Pi_i(h^t \oplus h^{T-t}(\sigma|_{h^t}))$. Had she not deviated her total payoff would have been $\Pi_i(h^{t-1} \oplus h^{T-t+1}(\sigma|_{h^{t-1}}))$. The threat against her deviation is *effective* if

$$\Pi_i(h^t \oplus h^{T-t}(\sigma|_{h^t})) \leq \Pi_i(h^{t-1} \oplus h^{T-t+1}(\sigma|_{h^{t-1}})) \ .$$

It can be shown that the profile σ is a subgame perfect equilibrium if and only if all threats against *one-player* deviations are credible and effective.[4] See Theorem 4.1 in [30] for the proof.

The following Lemma summarizes a few results about subgame perfect equilibria of repeated games that we use in the later chapters. Some illustrative examples are discussed in Chap. 4. A thorough discussion is given, e.g., in [32].

Lemma 2.1. *Let Q be the set of Nash equilibria of a stage game $\Delta = \langle N, A, \pi \rangle$. Let Δ^T be the game Δ repeated T-times.*

1. *A strategy profile that in every period, and regardless of history, selects some static equilibrium is subgame perfect. That is, if $\sigma^{t+1}(h^t) \in Q$ for any t and h^t then σ is subgame perfect.*
2. *If the stage game has a unique static equilibrium then the finitely repeated game has a unique subgame perfect equilibrium. It selects the static equilibrium in every period, regardless of the history. That is, if $Q = \{q\}$ then the unique subgame perfect equilibrium σ is defined by $\sigma^{t+1}(h^t) = q$ for any t and h^t.*
3. *If all static equilibria are payoff equivalent then all subgame perfect profiles are payoff equivalent. Any such profile selects in every period and regardless of the history one of the static equilibria. That is, if $\pi(q) = \pi(q')$ for any pair $q, q' \in Q$, then each subgame perfect equilibrium σ satisfies $\sigma^{t+1}(h^t) \in Q$ for any t and h^t.*

Sketch of the proof. (1) If in each period and independently of the history a static equilibrium is selected, then deviations in any period do not affect the continuation of the play. If she deviates the player therefore affects only her payoff in that period. However, since the play

[4] This is known as the one-stage-deviation principle.

in any period is a static equilibrium the player cannot increase her payoff with any deviation.

(2) This can be verified by the backwards induction argument as follows. In the last period a static equilibrium will be played. If the static equilibrium is unique, it is played in the last period regardless of the history. This implies that players' decisions in the penultimate period have no influence on their decisions and payoffs in the last period. Players should thus regard the penultimate period just as the last one and play the unique static equilibrium also in the penultimate period. By backwards induction, from the last period to the first, the same argument applies to all periods. This proves that whenever the stage game has a unique Nash equilibrium, the unique subgame perfect equilibrium of the finitely repeated game is to play the static equilibrium in each of the periods.

(3) Consider a repeated game with different static equilibria such that each player receives the same payoff in all of them. Consider a subgame perfect equilibrium of the finitely repeated game. In the last period some static equilibrium must be selected. The selection may depend on the history, but the players are indifferent about which of them is played. Their decisions in the penultimate period do not affect their payoffs in the last period, and the players regard the penultimate period just as the last one. A static equilibrium must thus be selected also in the penultimate period, and by backwards induction, in all the periods of the repeated game. Hence, an equilibrium path of any subgame-perfect equilibrium is characterized by a sequence of static equilibrium outcomes. □

2.2.3 Trigger Strategies

If all static equilibria are payoff-equivalent the subgame perfect equilibrium must select an equilibrium outcome in each period of the finitely repeated game. In contrast, it is shown in [33, 34] that if static equilibria are not payoff-equivalent then a subgame perfect equilibrium may be constructed whose equilibrium path is not a sequence of static equilibrium outcomes.

Effective and credible punishment threats may be constructed by making the selection of the static equilibrium that is played in the last periods conditional on the history of play in the preceding periods. It may be possible to punish a deviating player by playing, in the last period, the equilibrium which gives her the lowest payoff. On the other hand, a player that never deviated can be rewarded if her preferred equilibrium is selected.

Intuitively, in order to construct a credible threat for each player in each period, it is necessary that the set of static Nash equilibria consists of

- a rewarding equilibrium which is played (at least in the last period) if all players never deviated, and
- a set of punishment equilibria, one of which is played if some players deviate.

Each player must strictly prefer the rewarding equilibrium from her punishment equilibrium. Further static equilibria may exist which can be played if a group of players deviate simultaneously. For example, for each potential group of deviating players a static equilibrium may exist in which each of the deviating players earns less than in the rewarding equilibrium.[5] However, group punishment equilibria are not necessary for subgame perfection as long as players cannot coordinate their deviations.

Effective and credible threats may be constructed as follows. If no player deviates the reward equilibrium is repeated in the last periods. This generates the reward path. A deviation, however, triggers a repetition of the punishment equilibrium. This generates the punishment path. A repetition of some static equilibrium is always subgame perfect, hence the threats are credible. The effectiveness of a punishment path grows with its length. By repeating the punishment equilibrium sufficiently many times effective and credible threats may be constructed against any deviations.

The threats against deviations in the first periods of the game consist of long punishment paths. Toward the last periods of the game only shorter and therefore less effective punishment paths are possible. Because the effectiveness of threats decreases with periods, only the outcomes with less profitable deviations may be sustained in the last periods of the game.

Example 2.1 (Trigger strategies). Let q^* and q^0 be two Nash equilibria of the stage game $\Delta = \langle N, A, \pi \rangle$ such that $\pi_i(q^*) > \pi_i(q^0)$ for each player i. Consider also a non-equilibrium outcome $a \in A$ such that $\pi_i(a) > \pi_i(q^*)$ for each player i. All players prefer a over any of the two equilibria, and prefer q^* over q^0.

Even though a is not a static equilibrium it can be sustained in the subgame perfect equilibrium of the repeated game Δ^T. Consider, for

[5] For this, it is sufficient that one equilibrium is strictly preferred by all players to a particular another equilibrium.

example, the following strategy profile: "Play a during the early periods $1, .., T - t^*$. In case of a deviation jump immediately into the punishment equilibrium q^0 and play it until the last period. If no deviations occur during the early periods, play the reward equilibrium q^* in the remaining periods $T - t^* + 1, ..., T$."

The threats are credible because they consist of a repetition of a static equilibrium. Fix an early period $t \leq T - t^*$ and assume there were no deviations before t. If player i deviates at t she earns $(T - t)\pi_i(q^0)$ in the remaining periods. If she does not deviate at t or later she earns $(T - t - t^*)\pi_i(a) + t^*\pi_i(q^*)$. To make the threats during early periods effective the length of the reward phase t^* needs to be such that for any player i the smallest payoff loss from punishment, $t^*(\pi_i(q^*) - \pi_i(q^0))$, exceeds the most profitable one-period deviation from a. Such t^* exists and the corresponding strategy profile is subgame perfect for any $T > t^*$.

In the above example we outlined the simplest example of a trigger strategy which can sustain a non-equilibrium outcome in the early periods of a subgame perfect equilibrium. Any deviation from the non-equilibrium outcome in the early periods triggers an immediate switch to a punishment equilibrium, hence the name of the strategy. The following definition is due to Friedman [34].[6]

Definition 2.5. *Let Q be the set of Nash equilibria of a stage game $\Delta = \langle N, R, \pi \rangle$. Take $q^*, q^1, ..., q^n \in Q$, which possibly coincide, and a profile $r \in R$. A* ***trigger strategy profile*** *for Δ^T, based upon $(r, q^*, \{q^i\}_{i \in N}, t^*)$, $1 \leq t^* < T$, is denoted by $\sigma(r, q^*, \{q^i\}_{i \in N}, T, t^*)$ and is given by*

outcome path: play r during the early periods $1, ..., T - t^*$; play q^* during the remaining periods $T - t^* + 1, ..., T$,

threat: if player i deviates early, i.e. in an early period $t \in \{1, ..., T - t^*\}$, play q^i during the remaining periods $t + 1, ..., T$,

simultaneous deviations: if several players deviate in the same early period, exercise the threat for the deviating player with the lowest index.[7]

[6] Friedman [34] makes a distinction between trigger strategies and discriminating trigger strategies. In a trigger strategy profile each deviation triggers the same collective punishment. In a discriminating trigger strategy profile the punishments can be player specific. We do not emphasize this distinction and refer to all such profiles as trigger strategy profiles.

[7] The threat against simultaneous deviations is irrelevant for subgame perfection. Punishing the defector with the lowest index, however, assures that in any group

If no player ever deviates from the trigger strategy, the (possibly non-equilibrium) outcome r is selected in periods $1, .., T - t^*$, and the static equilibrium q^* is selected in periods $T - t^* + 1, ..., T$. Any deviation, on the other hand, triggers an immediate switch to the static equilibrium q^i, where i is the deviating player with the lowest index, selected ever after. A sufficient condition for the threat of punishment to be effective is that each player i strictly prefers q^* over q^i. The following theorem is due to Friedman [34].[8]

Theorem 2.1. *Let Q be the set of Nash equilibria of a stage game $\Delta = \langle N, R, \pi \rangle$. Take $q^*, q^1, ..., q^n \in Q$ and a profile $r \in R$ such that $\pi_i(r) \geq \pi_i(q^i)$ for each i. If $\pi_i(q^*) > \pi_i(q^i)$ for each player i then there exists a positive integer γ such that the trigger strategy profile $\sigma(r, q^*, \{q^i\}_{i \in N}, T, t^*)$ is a subgame perfect equilibrium of Δ^T for all T, t^* such that $T > t^* \geq \gamma$.*

2.2.4 Prisoner's Dilemma

Prisoner's dilemma games stylize situations in which there is a conflict between individual and common interests. Consider a pair of crime suspects being investigated in separate locations and with no possibility of communication. Each is given an opportunity to witness against the other. The court considers any witness favorably. If both confess each receives a sentence of 3 years in prison. If only one confesses against the other, he is set free and the other receives a sentence of 4 years. If neither confesses each receives a sentence of 1 year for a minor offense.

The suspects achieve the mutually optimal outcome if neither confesses. However, each has an incentive to confess regardless of the behavior of the other. Hence, rational suspects always witness against each other and serve 3 years each.

Let v, the payoff function of the prisoner's dilemma game $\langle \{1, 2\}, \{C, D\}^2, v \rangle$, be described by the following payoff matrix,

		Player j C	Player j D
Player i	C	c, c	e, f
	D	f, e	d, d

(2.1)

of potential defectors one player will not want to participate in a simultaneous defection.

[8] The original theorem in [34] allows for player-specific discount rates. Throughout this chapter we assume, for simplicity, that players do not discount future. We thus state a simplified version of the original theorem.

where $f > c > d > e$, and $2c > f+e$. We interpret action D as defection and action C as cooperation. An instance of a prisoner's dilemma game is shown in Fig. 2.1 on p. 9. For each player action D is the best response to any action of the other player. Hence, (D, D) is the unique Nash equilibrium of a prisoner's dilemma game.

Several generalizations of the prisoner's dilemma game can be made. An n-player prisoner's dilemma is the game $\langle N, \{C, D\}^n, \pi \rangle$, where

$$\pi_i(a) = \sum_{j \in N \setminus \{i\}} v_i(a_i, a_j) .$$

In the n-player prisoner's dilemma each player chooses either cooperation or defection and plays a two-player game with each other player in turn.[9]

In each of the prisoner's dilemma games above defection is the unique best response of each player. Hence, each of the games above has a unique Nash equilibrium in which all players defect. In contrast, the collectively optimal outcome in each of these games is attained when each player cooperates. The individual incentives thus conflict with common interest. This conflict is cumulated if any of the prisoner's dilemma games above is finitely repeated: there is no cooperation in the unique subgame perfect equilibrium (see Lemma 2.1(2)).

A common feature of all these games is the fixed group structure: players do not choose with whom they play the game. In this book we explore the possibility to attain cooperation if players are free to choose their partners and establish the group structure endogenously. The corresponding stage game will be introduced in Chap. 3. We outline in the following section the basic concepts of the mathematical theory of networks.

2.3 Elements of Network Theory

A network consists of a set of agents and a set of links. Each link connects a pair of agents. Formally, a network can be described by the mathematical concept of a graph. The elements of the network theory, introduced in this section, coincide with several basic concepts of graph theory. For advanced reading in the mathematical theory of graphs see e.g. [36].

[9] This game can also be described as a linear public goods game with discrete binary choice, see [35].

We consider networks in which interaction is undirected. Whenever two agents have a mutual link they both interact with each other. We do not discuss directed networks and thus omit the classifier "undirected".

Definition 2.6. *A **network** is characterized by*

- *a finite set of **agents** $N = \{1, ..., n\}$, and*
- *a set of **links** $L = \{ij \mid \{i, j\} \in N \times N$ and $i \neq j\}$.*

Following the mathematical convention we denote the *link* between agents i and j simply by ij. In the definition of a network we implicitly assume that links are binary: two agents are either linked or not. Furthermore, a pair of agents can have at most one link and no agent can have a link with herself.

A network (N, L) can be represented by an $n \times n$ binary *adjacency matrix* g, where

$$g_{ij} = 1 \text{ if } ij \in L \text{ and } g_{ij} = 0 \text{ otherwise} .$$

Matrix g is symmetric: if $ij \in L$ then $ji \in L$. We refer to g simply as network.

Agents i and j are *neighbors* in network g if they have a mutual link, that is, if $g_{ij} = 1$. Set $L_i = \{j \mid ij \in L\}$ is the *neighborhood* of agent i. The size of i's neighborhood is the number of her links $l_i = |L_i|$.[10] Agent i is *isolated* if it has no neighbors, i.e. if $l_i = 0$.

A *path* of length k between agents i and j is the sequence of agents $(i, i_1, ..., i_{k-1}, j)$ such that $g_{ii_1} = g_{i_1 i_2} = ... = g_{i_{k-1}j} = 1$. The (geodesic) *distance* between i and j is the length of the shortest path between them. The distance between i and j is infinite if there is no such path, and 1 if i and j are neighbors. If i and j are not neighbors but there is a path between them we say that they are *indirectly linked.*

A graph is *connected* if every pair of agents is (indirectly) linked. If a graph is not connected it has several connected components. A *component* of g is a subset of agents $N' \subseteq N$ such that (i) all agents in N' are (indirectly) linked and (ii) agents in N' have no links with agents in $N \backslash N'$. An isolated agent, for example, constitutes a component herself. A subset of agents $N' \subseteq N$ is a *clique* if all agents in N' are neighbors, that is, if each agent in N' is linked to every other agent in N'.

A large body of literature on networks deals with network measures. A network cannot be captured by one unidimensional measure. Analysis of social, economic or technological networks usually relies on several such measures, each designed to capture one particular aspect of the

[10] In graph theory this is the degree of node i.

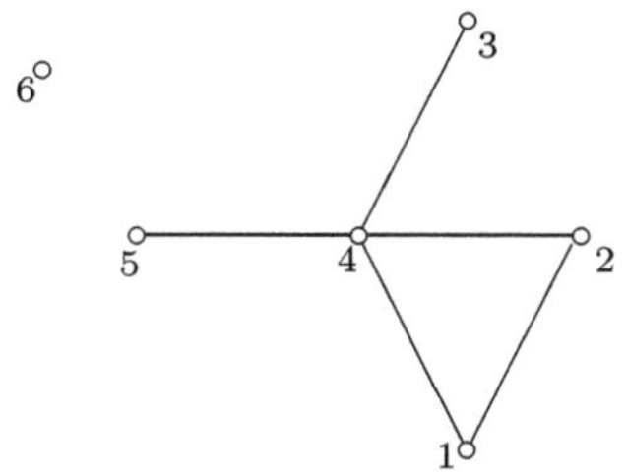

Fig. 2.2. A network with 6 agents

network structure of topic-specific significance. Social scientists, for example, study network centrality, power distribution and cohesion, and have developed several measures of each (see e.g. [37, 38]). Economists are interested in efficiency and equality of networks, and engineers care for network reliability and throughput. We evoke one simple measure that is used in most network analyzes. A *density* of a network is the number of established links divided by the number of all possible links,

$$density(g) = \frac{2\,|L|}{n(n-1)} .$$

Example 2.2. For demonstration of several concepts we discuss the network illustrated in Fig. 2.2. Circles correspond to agents and lines correspond to links. For example, there is a link between agents 1 and 2. The network has 6 agents and 5 links. The neighborhood of agent 1 consists of agents 2 and 4. Agent 1 is also indirectly linked to agents 3 and 5, but not to agent 6. Two different paths connect agent 1 with agent 5: (1,4,5) and (1,2,4,5). The shortest one is of length two, which is then the distance between agents 1 and 5. The agents {1,2,3,4,5} constitute a connected component, but the network is not connected. Player 6 is isolated. Players 1, 2 and 4 constitute the largest clique in the network. This network can be captured by a 6×6 binary matrix g, defined by $g_{12} = g_{21} = g_{14} = g_{41} = g_{24} = g_{42} = g_{34} = g_{43} = g_{45} = g_{54} = 1$ and otherwise $g_{ij} = 0$. Its density is $\frac{1}{3}$.

Several classes of networks, called *network architectures*, are prominent in the network literature. Figure 2.3 illustrates the following four architectures.

Empty network No links are established: $g_{ij} = 0$ for all i and j.
Complete network All links are established: $g_{ij} = 1$ for all i and j.
Wheel network Links constitute a cycle spanning all agents. There is an ordering of all agents $(i_1, i_2, ..., i_n)$ such that $g_{i_1 i_n} = g_{i_n i_1} = 1$ and $g_{i_j i_{j+1}} = g_{i_{j+1} i_j} = 1$ for all $j \in N \backslash \{n\}$ and $g_{k,k'} = 0$ otherwise.

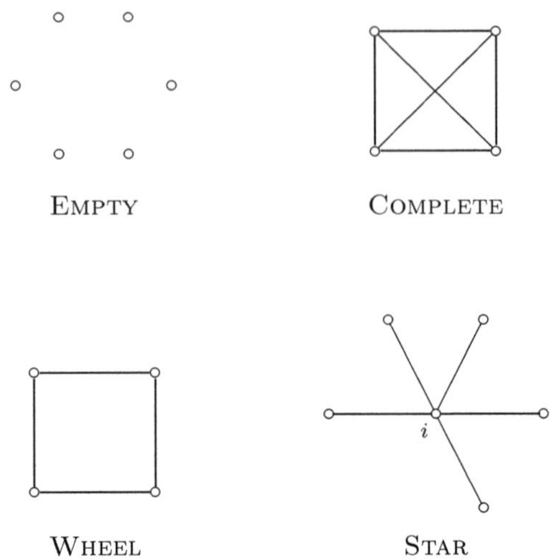

Fig. 2.3. Common network architectures

Star network There is one central agent, linked to all other agents. There are no other links. If i is the central agent then $g_{ik} = 1$ for all $k \in N \backslash \{i\}$ and $g_{jk} = 0$ for all $j, k \in N \backslash \{i\}$.

For convenience we introduce the following notation. Given a network g and a pair of distinct agents ij let

$$g \oplus ij$$

be the network established by adding link ij to network g, and let

$$g \ominus ij$$

be the network established when link ij is removed from network g. If ij is already present in g then $g \oplus ij = g$. Similarly, if ij is not present in g then $g \ominus ij = g$.

2.4 Elements of the Theory of Markov Chains

In the following we shortly describe the elements of the theory of stationary discrete-time Markov chains. The theory is used in the analysis of dynamic processes in Chap. 6. An accessible introduction into Markov chain theory may be found in [39]. Further details and proofs may be found in advanced textbooks on the theory of stochastic processes, e.g. [40].

Markov chains are among the simplest stochastic processes. A *stochastic process* is, roughly, an infinite sequence of random events. The sequence often represents the ordering of events in time, in which case we can speak of past and future events, of a current state, and of a history. Such a stochastic process is described by a set of possible states, and by a function that assigns a probability distribution over the future events to any history.

A stochastic process is *discrete-time* if events occur only at given, discrete points in time, usually denoted by $t = 0, 1, 2, \ldots$. In such a process an event describes a transition from one possible state to another. At each discrete time point t, the *history* of the process describes the sequence of states that the process went through at times $0, \ldots, t$. A discrete-time stochastic process is described by a set of possible states and, for each possible history, a probability distribution over possible transitions between the states. The probability distribution over possible transitions between times t and $t+1$ may, in general, depend on the whole history up to t. If, however, at each time point it depends only on the current state then the process is a Markov chain.

A discrete-time *Markov chain* is therefore a random sequence of transitions between states, such that at each time t the probability distribution over transitions between times t and $t+1$ is independent of the past states of the system during times $0, \ldots, t-1$, but may depend on the current state at t. A Markov chain is *stationary* if the distribution over future transitions is independent of the time *index* t. For simplicity we omit general definitions and define only the Markov chains used in this book.

Definition 2.7. *A finite stationary discrete-time* ***Markov chain*** *is characterized by*

- *a finite* ***state space*** $S = \{s_1, \ldots, s_k\}$ *and*
- *a* $k \times k$ *stochastic* ***transition matrix*** M. [11]

The Markov chain starts at time 0 in some initial state $s^0 \in S$. The choice of the initial state can be random. The evolution of the process continues as follows. Let $s[t]$ be the state of Markov chain at time t. If at time t the Markov chain is in state s then the probability that it will be in state s' at time $t+1$ is

[11] A $k \times k$ matrix M is *stochastic* if each row constitutes a probability vector, that is, if $M_{ij} \geq 0$ for each i, j and $\sum_{j=1}^{k} M_{ij} = 1$ for each row i.

$$M_{ss'} = prob[s[t+1] = s' \mid s[t] = s] \; .$$

The value of $M_{ss'}$ therefore gives the probability of the *transition* from state s into state s'. The value of M_{ss} is the probability that the process will not make any transition from state s.

For any non-negative integer τ let M^τ be the τ-fold product of the matrix M, that is

$$M^\tau = \underbrace{M \cdot M \cdot \ldots \cdot M}_{\tau} \; .$$

For each $t, \tau \in \mathbb{N}$ the value of $M^\tau_{ss'}$ gives the probability that the process is in state s' at time $t + \tau$, conditional on being in state s at time t.[12] A state s' is *accessible* from state s if there exists a finite $\tau > 0$ such that $M^\tau_{ss'} > 0$. The process is *aperiodic* if for every pair of states s and s' such that $M^\tau_{ss'} > 0$ there exists a finite $t > 0$ such that $M^\tau_{ss'} > 0$ for all $\tau \geq t$.

A set of states $Z \subseteq S$ is *recurrent*, if (i) every state $s' \in Z$ is accessible from every other state $s \in Z$, and (ii) the states outside Z are not accessible from the states in Z: if $s \in Z$ and $s'' \in S \backslash Z$ then s'' is not accessible from s.

It can be shown that every Markov chain with a finite state space has at least one recurrent set. A state $s \in S$ is an *absorbing state* if $\{s\}$ constitutes a singleton recurrent set. If the process reaches an absorbing state it remains in that state forever. A state that is not absorbing and does not belong to a recurrent set is called *transient* . Every Markov chain eventually leaves transient states and converges to a recurrent set with probability one. For the analysis of the long run behavior of a Markov chain it is therefore necessary to characterize the recurrent sets of the process.

Suppose an initial state $s^0 \in S$. For each $t > 0$, let $\mu^t(s \mid s^0)$ be the relative frequency with which state s is visited during the first t periods. As t goes to infinity, $\mu^t(s \mid s^0)$ converges almost surely to a probability distribution $\mu^\infty(s \mid s^0)$, called the *asymptotic frequency distribution* of the process conditional on s^0. Distribution μ^∞ can be interpreted as a selection criterion: over the long run, the process selects only those states on which $\mu^\infty(s \mid s^0)$ puts positive probability. Which states are selected in this sense may depend on the initial state s^0. If it does not, that is, if the asymptotic distribution $\mu^\infty(s \mid s^0)$ is independent of s^0, we say that the process is *ergodic*.

[12] In matrix terms, if ϱ is the probability vector over the states at time t, the probability vector at time $t + \tau$ is given by ϱM^τ.

Consequently, for every initial state s^0 the states on which $\mu^\infty(s \mid s^0)$ puts positive probability constitute a union of recurrent sets. Hence, the process is ergodic if and only if it has a unique recurrent set.

The asymptotic properties of finite Markov chains can be studied as follows. Let $\mu = (\mu(s))_{s \in S}$ be a probability distribution on S. Consider the system of linear equations

$$\mu M = \mu, \text{ where } \mu \geq 0 \text{ and } \sum_{s \in S} \mu(s) = 1 . \tag{2.2}$$

It can be shown that this system always has at least one solution μ, which is called a *stationary distribution* of the Markov chain with transition matrix M. Equation (2.2) states that, if the probability distribution over the initial states is given by μ, the probability distribution over the states of the process at any time $t > 0$ is also given by μ. The dynamics of the process is in this sense stationary, which motivates the use of the term.[13] In general a Markov chain may have many stationary distributions. If, however, the process is ergodic the stationary distribution is unique and, furthermore, coincides with the asymptotic frequency distribution $\mu^\infty(s \mid s^0)$ for any s^0. More precisely, for an ergodic process,

$$\lim_{t \to \infty} \mu^t(s \mid s^0) = \mu^\infty(s \mid s^0) = \mu(s) \text{ for all } s, s^0 . \tag{2.3}$$

The following proposition illustrates these ideas for a Markov chain in which transition between any pair of states occurs with positive probability. These results are used in Chap. 6. Proofs may be found in [39].

Proposition 2.1. *Consider a transition matrix M^* with only positive entries: $M^*_{ss'} > 0$ for all $s, s' \in S$. The corresponding Markov chain has a number of properties:*

1. *the unique recurrent set consists of the whole state space S,*
2. *the process is aperiodic and ergodic,*
3. *the stationary distribution $\mu(s)$ is unique,*
4. *regardless of the initial state, for large t the probability that the process is in state s is approximately $\mu(s)$, and*
5. *regardless of the initial state, the proportion of time that the process spends in state s is approximately $\mu(s)$.*

---

[13] Above we also refer to stationary Markov chains, which should not be confused with a stationary distribution of a Markov chain.

3

The Network Dilemma Game

In this chapter we introduce and discuss our model of a social dilemma with endogenous choice of partners: the network dilemma game. All subsequent chapters deal with this game or its variations. In Sect. 3.1 we introduce notation and define the basic game. In Sect. 3.2 we discuss some related examples and discuss what real-life features we have omitted in the zeal for clarity and parsimony. We conclude with a brief review of the related literature.

3.1 Definition

We consider an n-person prisoner's dilemma game with endogenous partner choice. In this game each of the n players simultaneously makes two decisions:

LINKING DECISION: Each player proposes links between herself and other players. Linking is costless. Each player can propose a link to any other player in the group. Proposing a link may or may not be sufficient to establish it. When a link between two players is established we say the players are neighbors.

CHOICE OF ACTION: Each player chooses an action in a prisoner's dilemma game. One game is played by each pair of neighbors. Players, however, cannot discriminate in their action choices. They can either cooperate with all their neighbors or defect on all of them. A player receives an outside option for each player she is not linked with.

The payoff to a player is the sum of the earnings from the prisoner's dilemma games played with her neighbors and of the outside options received for not playing with the remaining players.

At most one link can be established between each pair of players. We say that linking is unconstrained when each player is free to propose and establish any number of links, up to $n-1$. Otherwise we say that linking is constrained. We introduce here, for the game with unconstrained linking, the notation used throughout all chapters. This game is a special case of the game with linking constraints, which we define and analyze in Chap. 4.

To distinguish between different kinds of choices we use the following notation:

- a *linking choice* describes the links proposed by a player,
- an *action* refers to a player's choice between cooperation and defection,
- a *move* consists of a linking choice and an action and thus describes all choices made by the player in the game described above.

Let $N = \{1, ..., n\}$ be the set of players. The linking choice of player i can be captured by a binary vector

$$p_i = (p_{ij})_{j \in N} \in \{0,1\}^n, \text{ such that } p_{ii} = 0 \ .$$

If player i proposed a link to player j then $p_{ij} = 1$, otherwise $p_{ij} = 0$. The constraint $p_{ii} = 0$ is assumed for convenience, and indicates that a player cannot establish a link with herself. If $p_{ij} = 0$ we say that player i *refused to link* to player j. Linking choice p_i is *trivial* if $p_{ij} = 0$ for all j, that is, when no links are proposed by player i.

For each profile of linking choices $p = (p_1, ..., p_n)$ let $g(p)$ denote the corresponding *network of established links*. Two models of link formation are common in the literature [41]. Each describes how the network is established from a profile of proposed links. The two models of link formation are formalized as follows:

MUTUAL LINK FORMATION: A link between players i and j is established when $p_{ij} = 1$ and $p_{ji} = 1$. That is, $g_{ij} = \min\{p_{ij}, p_{ji}\}$.

UNILATERAL LINK FORMATION: A link between players i and j is established when $p_{ij} = 1$ or $p_{ji} = 1$. That is, $g_{ij} = \max\{p_{ij}, p_{ji}\}$.

In the mutual linking model consent of both players is needed to establish a mutual link. That is, a link between two players is established if and only if it is proposed by both players. In the unilateral linking model no second party consent is needed to establish a link. A link between two players is then established whenever it is proposed by at least one of them. In this book we focus on mutual link formation, as it is the more realistic description of formation of social connections.

See [42] for an investigation of cooperation in networks formed through unilateral linking.

Whenever a link between two players is established each of them interacts with the other, regardless of how the link was established and by whom it was proposed. For each network g let $L_i(g) = \{j \mid g_{ij} = 1\}$ be the set neighbors of player i and let $l_i(g) = |L_i(g)|$ be the size of her neighborhood. We say that two players without an established link are *separated*, that is, they are not neighbors. For convenience we use the shorthand notation $L_i(p) \equiv L_i(g(p))$ and $l_i(p) \equiv l_i(g(p))$ for a profile of linking choices p.

The action of player i in the prisoner's dilemma game is denoted by $a_i \in \{C, D\}$. Let $v(a_i, a_j)$ denote the payoff to player i choosing action a_i for playing the game with player j choosing action a_j, where the payoff function v is given by the following payoff matrix,

		Player j C	D
Player i	C	c, c	e, f
	D	f, e	d, d

where $f > c > d > e$ and $2c > e + f$. As mentioned above, player i plays a prisoner's dilemma game with all her neighbors and receives an *outside option* for each other player that is not her neighbor. Let $o \in \mathbb{R}$ be the value of the outside option. We distinguish between

high outside option: $o > d$, the outside option is more valuable than mutual defection; and
low outside option: $o < d$, the outside option is less valuable than mutual defection.

A *move* of player i is a pair (a_i, p_i), where a_i is her action and p_i her linking choice. Let $A_i = \{C, D\}$ be the set of actions and let $P_i = \{p_i \in \{0,1\}^n \mid p_{ii} = 0\}$ be the set of linking choices of player i. The *set of moves* of player i is denoted by $J_i = A_i \times P_i$. Let $J = \times_{i \in N} J_i$. A *profile of moves* is the n tuple of pairs denoted by $(a, p) \in J$. The components of a profile of moves, a and p, are the *action profile* and the *linking profile*, respectively.

Let (a_{-i}, p_{-i}) denote a profile of moves of player i's opponents $N \backslash \{i\}$, and let $J_{-i} = \times_{j \in N \backslash \{i\}} J_j$ be the set of all such profiles. For each profile $(a, p) \in J$ the payoff to player i is given by her payoff function $\pi_i : J \to \mathbb{R}$, defined by

$$\pi_i(a,p) = \sum_{j \in L_i(p)} v(a_i, a_j) + (n - 1 - l_i(p))o \; . \tag{3.1}$$

We also define $\pi : J \rightarrow \mathbb{R}^n$ as the function whose i-th component is π_i.

The set $L_i(p)$ and its size $l_i(p)$ by definition depend only on the links that are established. By implication, for a given action profile a the payoff $\pi_i(a,p)$ of player i also depends only on the network of established links $g(p)$. With this in mind we may slightly abuse notation and define $\pi_i(a,g)$ for any network g by

$$\pi_i(a,g) = \sum_{j \in L_i(g)} v(a_i, a_j) + (n - 1 - l_i(g))o \; . \tag{3.2}$$

We refer to the stage game $\Gamma = \langle N, J, \pi \rangle$ as the *network dilemma game with unconstrained linking*, or shortly, the *basic game.*

Two particular profiles of actions are convenient to define. Let $\mathbf{D} = (D, ..., D)$ and $\mathbf{C} = (C, ..., C)$ denote the profiles of uniform defection and of uniform cooperation, respectively.

An established link between two cooperative players represents a *cooperative relation.* An established link between a cooperative and a defective player represents a *semi-cooperative relation.* An established link between two defective players represents a *defective relation.*

3.2 Discussion

The network dilemma game is a simple, abstract, but fundamental model of a situation where agents are free to select their interaction partners and face an inherent free-rider problem in their interactions. The game-theoretic models of social dilemmas traditionally assume that participation in the game is non-voluntary. However, in many real situations, for example in human social networks, agents are free to choose whether or not and with whom to interact. It is, therefore, important to develop and study models of interaction in endogenous networks. Of course, our game is an abstraction from reality and might not be an accurate representation of all related real-life situations. It does, however, capture the essential features of many such situations better than models of social dilemmas that assume an exogenously imposed interaction structure.

Few other game-theoretic models of interactive situations with endogenous interaction structures exist in the literature. Most focus on games of coordination in endogenous networks which are similar to our network dilemma game except that interaction is modeled as a coordination game instead of the prisoner's dilemma game [43, 44, 45]. A slight variation of these models is based on an assumption that links change relatively less frequently than behavior [46]. The multiple prisoner's dilemma games with outside option in [25] and [24] differ from our network dilemma game in that discrimination in actions is possible across neighbors. This essentially reduces the game to a collection of two-player prisoner's dilemma games with outside option, with little or no significance of the network structure. A common variation of this model considers a two-phase process of link formation: in the first phase each player proposes one link and then decides which of the proposed links to accept in the second phase [47, 48]. A related game in an evolutionary setting is studied in [49]. We allude to these models in more detail in subsequent chapters. In Chap. 4, for example, we indicate that our theoretical results hold regardless of whether or not discrimination in actions across the neighbors is possible.

There are some real-life situations that are captured well by our network dilemma game, such as for example the exchange networks in Papua New Guinea and Ghana, see [9, 50]. These are instances of voluntary exchange networks with incomplete contracts. For an illustration of such a network consider a village exchange economy that consists of a number of villagers each uniquely specialized in producing one good. A villager can produce either low or high quality goods and can obtain other goods through bilateral exchange. Producing goods of different quality requires different tools and investments which makes simultaneous production of goods of different quality unfeasible for an individual villager. No villager can be forced to produce goods of high quality but it is possible to refuse bilateral exchange with another villager. The network dilemma game captures this situation if we assume that villagers do not know the quality of other villager's goods before the exchange. Contemporary instances of voluntary exchange networks with incomplete contracts can be found in production networks among specialized firms whose production processes require advance investments that permit only periodic adjustments of quality, and in markets for "experience goods" [51]. On a larger scale, the international trade network may be viewed as another such instance: we may interpret the players as countries, defection as the decision of a country to subsidize its economy, and exclusion as the imposition of tariffs or quotas for goods imported from another country.

Peer-to-peer computer file-sharing networks can also be seen as situations that carry features of a network dilemma game. A file sharing network consists of a (usually large) number of members, each able to share some of her computer files with other members by making them freely available for download. A typical peer-to-peer network faces an inherent free-rider problem: each member benefits from the increased number of files shared by the other members but prefers not to share her own files, for example because of the resulting copyright violation. To counter this problem the file-sharing services typically provide public classification of its members according to their past or current volume of shared files and permit each member to use this classification to bar selected other members from downloading her shared files. Recent research suggests that selective choice of file-sharing partners based on their reputation increases the volume of sharing [52] and that free-riding may be reduced even if information about reputation is decentralized and flows through bilateral links between the file-sharing partners [53]. The volume of sharing may be further increased if exclusion of free-riders is enforced by the file-sharing service itself [54].

There are, of course, limits to the applicability of our network dilemma game for exploration of real-life situations. For example, our game cannot capture the sequential nature of decisions in human social networks, where the possibilities to serve or help occur at different times for different persons. Moreover, the real process of formation of social links is more complex than presumed by our simple network formation models. Finally, in contrast to many real-life situations, but in common with virtually all similar theoretical models, we assume that all players are homogeneous (see [55, 56, 57] for models of formation of networks among heterogeneous players). We intend to address these points in future research by tailoring the models to capture specific social networks, such as networks of informal insurance.

We conclude this chapter with a brief overview of models of network formation in the game-theoretic literature (see [41] for a more thorough discussion). The mutual and unilateral link formation models were introduced in [27] and [58], respectively. The mutual link formation model seems more realistic and is more frequently applied. Unfortunately, it is less suitable for analysis with off-the-shelf game theoretic tools because it induces a problem of multiplicity of equilibria. We discuss this in detail in Chap. 4.

Jackson and Wolinsky [59] avoid the problem of multiple equilibria by investigating the stability of established networks rather than the equilibria of the network formation game. They define a network to be *pairwise-stable* if no player can benefit by removing one of her links and there is no pair of *separated* players such that each of them benefits if they establish the mutual link. The following adaptive dynamic process always converges to a pairwise-stable network: at each time two players are selected at random and myopically decide whether to add or remove their link, or to make no changes [60]. Jackson and Watts [61] introduce perturbations to this process and use the criterion of stochastic stability to identify the most robust among the pairwise-stable networks. Several related results are reviewed and synthesized in [41]. Interesting extensions of the concept of pairwise-stability are discussed in [62].

Aumann and Myerson [63] model network formation as a finite game in extensive form. The game is described by an exogenously given sequence over all possible pairs of players. In order of this sequence two players are selected in each period who then decide whether or not to establish the mutual link, having perfect foresight about the links that will be established in subsequent turns. For variants of this game see [64, 65, 66, 67]. The drawbacks of this approach are that this game is difficult to analyze even in the one-shot setting and that the set of equilibria is sensitive to the choice of the initial sequence.

A related line of research deals with cooperative games that have communication structures. It originated with coalition formation models in [68] and [69] and assumes that payoffs depend only on which players are directly or indirectly connected, but not on how they are connected. This literature is surveyed in [70]. Such an approach is not really suitable for modeling network formation because it neglects the effects of the structure of networks (such effects can be introduced through linking costs, see [71]).

Finally, a distinct approach is taken by Skyrms and Pemantle [72] who model network links as continuous rather than binary variables. In their model a link is characterized with the probability that the two involved players will interact in any given period. They investigate the network dynamics if the probability of future interactions depends on the rewards from past interactions.

4

Cooperation in Finitely Repeated Network Dilemma Games

Modeling an interactive situation as a non-cooperative game is perhaps the most common approach in game theory. The standard assumption is that players are rational and have a perfect foresight. If they are not perfectly informed about the situation or about the past behavior of other players they still know all the relevant chances. In addition, preferences and information of each player are common knowledge to all of them. These assumptions facilitate a relatively straightforward analysis that often provides simple and exact results. The conventional task is to find the equilibria of the game. An equilibrium is any collection of actions, one for each player, such that it is in each player's self interest to play her action if all other players play theirs.

A well known example is that of a finitely repeated prisoner's dilemma game which, with rational and perfectly informed players, has a unique equilibrium. In this equilibrium all players mutually defect in all periods even though they know that mutual cooperation would have been better for each of them. In this chapter we contrast this result by showing that cooperation among such players may yet be achieved when they can choose their partners endogenously. We do this by exploring the subgame-perfect equilibria of finitely repeated network dilemma games. Our focus is on the effect of linking constraints for cooperation. Our intuition is that competition for partners, induced by scarcity of links, may induce rational players to cooperate. This is confirmed by our results.

In the following section we motivate our approach, outline the results and discuss our contribution in view of the related literature. We introduce and formally define the network dilemma games with linking constraints in Sec. 4.2 and characterize their equilibria in Secs. 4.3 and 4.4. A discussion of our results is given in Sec. 4.5, where we also suggest some topics for future investigation. We summarize the chapter in Sec. 4.6.

4.1 Introduction

The prisoner's dilemma game is a paradigm used to study the failure to achieve cooperation in groups of rational and informed players. The finitely repeated prisoner's dilemma, the multi-player prisoner's dilemma and the prisoner's dilemma played on a fixed network are examples of games where the unique equilibrium consists of strategies that prescribe defection in each of the periods even though mutual cooperation benefits each of the players. Cooperative equilibria exist only in infinitely repeated prisoner's dilemma games or in those with random ending (see the Folk theorem and its extensions in [73, 13, 74]).

To achieve cooperation in the finitely repeated prisoner's dilemma the assumptions about players and their environment must be relaxed (see [14] for relaxation about information and [75] for relaxations about rationality). In this chapter we investigate whether cooperation can be achieved in equilibrium under standard assumptions, if we change the game by permitting the players to choose their partners.

We consider network dilemma games with mutual link formation. Traditional equilibrium concepts from non-cooperative game theory, such as Nash equilibrium and subgame-perfection, can straightforwardly be applied to network formation games with unilateral linking [58]. However, these concepts may be too weak for network dilemma games with mutual linking. In particular, in such games a large number of unintuitive Nash equilibria exist simply because of mis-coordination where pairs of players want to establish a link but neither proposes it, because neither expects the other player to do so. In our investigation of repeated network dilemma games with mutual linking this multiplicity of Nash equilibria further translates into the existence of unintuitive subgame-perfect equilibria. To avoid the problem of multiplicity of equilibria in network formation games the requirement of Nash equilibrium is usually supplemented with that of network stability. This approach has been formalized by the concept of a pairwise-stable equilibrium and its variants [76, 77, 78]. However, these equilibrium concepts are defined only for network formation games with *exogenous* linking costs and benefits. They cannot be applied to network dilemma games, where benefits of links depend on actions taken by players in the prisoner's dilemma. In this chapter we define the concept of a *linking-proof equilibrium* which may be seen as an extension of the concept of pairwise-stable equilibrium to network formation games with endogenous benefits. For repeated games with mutual linking we define a related concept of a *linking-proof subgame-perfect equilibrium.*

In our analysis we focus on network dilemma games with linking constraints. In the network dilemma games defined in the previous chapter players are free to establish any number of links. In many real situations, however, agents are strictly constrained in the maximal number of connections they can maintain. One such example is the human social network: people select only a few of the other people with whom they keep regular contacts. The number of contacts is obviously negligible in comparison to the number of all other people but typically, and more crucially, it is small also compared to the number of people in some reference group, such as school class or job environment.

Constraining the number of links that players can propose or establish has a dramatic effect for cooperation in network dilemma games with low outside option. We say that a player's linking constraint is strict when she cannot simultaneously establish links to all other players. In this chapter we show that cooperation can be achieved in a (linking-proof) subgame-perfect equilibrium as long as at least two players have a strict linking constraint. This condition is relatively weak and not satisfied only in a few boundary cases, such as by the original network dilemma game without any linking constraints. We proceed to identify network dilemma games in which cooperation can be achieved using only simple and intuitive trigger strategies that threaten to punish any deviation with exclusion. Such a strategy profile exists only when there is a sufficient variety of (linking-proof) equilibrium networks, so that the group can effectively punish any player by removing some or all of her established links and still structure into an equilibrium network. We show that this is possible if linking constraints are relatively severe, that is, if each player's upper bound on the number of her neighbors is relatively small. The human social network is again a suitable example.

The intuition behind our results is as follows. When linking is strictly constrained and the outside option is low the players can structure their interaction into several different *equilibrium* networks. Because the outside option is low, each player prefers to establish as many links as possible. In some equilibrium networks all players establish their maximal number of links, but in others some players establish less or even no links. We could say that some equilibrium networks are efficient and others are inefficient. Naturally, in the final period of a finitely repeated network dilemma game all rational players defect. However, because their interaction in the final period can be structured into a variety of networks they can condition the structure on the actions taken in the previous periods. To see this, suppose that players begin by cooperating in an efficient network. If one of them defects the remaining players can punish her by structuring their final period interactions into an inefficient equilibrium network in which this player establishes relatively little or no links, earning only the low outside option. This player is thus punished by exclusion. When trigger strategies fail to sustain cooperation we may still use, slightly less simple, recursive trigger strategies. For this it is sufficient that at least one player can be excluded from a connected equilibrium network. This, in turn, is possible with almost any linking constraints.

We can relate our results to the existing literature about subgame-perfect equilibria in finitely repeated games. Friedman [34] and Benoit and Krishna [33] have proven a limit Folk theorem for finitely repeated games when the set of static equilibrium payoffs has sufficient dimensionality. Within network dilemma games, a large variety of equilibrium networks implies high dimensionality of static equilibrium payoffs. It follows from Friedman's result that this may be sufficient to sustain non-equilibrium outcomes, such as cooperation, in a subgame-perfect equilibrium. Furthermore, Smith [79] has proven the limit Folk theorem for finitely repeated games under relaxed conditions on dimensionality of static equilibrium payoffs. Our characterization of when the recursive trigger strategies can sustain cooperation in network dilemma games is related to his condition of recursively distinct Nash payoffs. For a review of related results in a unified framework see [80].

We contribute to this literature by highlighting the possibility that the dimensionality of equilibrium payoffs can be increased by endogenizing the interaction structure. When the interaction structure is fixed then in some games, such as the prisoner's dilemma, only the static equilibrium outcomes can be sustained in a subgame-perfect equilibrium. We demonstrate that this may change if the interaction network is endogenous: outcomes other than static equilibria may be achieved in a subgame-perfect equilibrium of the network dilemma game under any of the two models of link formation. Furthermore, given that the concept of a subgame-perfect equilibrium may be too weak for games with mutual linking we show that, when linking is constrained, the above results hold also under the additional requirement of linking-proofness.

Our study is one of relatively few that consider situations where agents form the network and also determine its value through behavior in the network interactions. Most studies investigate games of coordination in endogenous networks, assuming either unilateral or mutual link formation [43, 44, 45]. These games may be viewed as equivalent to our network dilemma games for situations where interaction over the network has a character of a coordination game instead of the prisoner's dilemma game. These studies typically consider an adaptive dynamic process with myopic best-responding players, and study the long-run stability of different conventions in relation to the linking costs. One somewhat surprising conclusion is that high linking costs imply high earnings and efficiency.

Evolutionary dynamics is the preferred approach also in studies of social dilemmas on endogenous interaction structures. These studies assume that behavior spreads via imitation rather than the best response, ostensibly because defection is the unique best response in social dilemma situations. We review this literature in Chap. 6, where we also show that cooperation can be sustained in network dilemma games even with myopic best-responding players, if they use limited forward looking. An equilibrium analysis of an infinitely repeated prisoner's dilemma game, played on an endogenous network by rational players with perfect foresight, is given in [81]. In this chapter we consider *finitely repeated* network dilemma games and analyze whether cooperation can be sustained in a subgame-perfect equilibrium through strategic linking behavior.

We are aware of only two other studies that study strategic behavior in similar situations. Hirschleifer and Rasmusen [82] study a finitely repeated n-player prisoner's dilemma game in which each period of the prisoner's dilemma is followed by a vote whether to ostracize any players from the group. Cooperation is shown to be possible in the initial periods of a subgame-perfect equilibrium if ostracism in the last period of the game is costless for ostracizers and costly for the ostracized. Their model does not deal with individual exclusion but assumes that each individual conforms to the outcome of the vote.

The other related result is about finitely repeated multiple prisoner's dilemma games with (high) outside option, analyzed by Hauk and Nagel [24]. They show that cooperation can be achieved in a Nash equilibrium of the repeated game if link formation is unilateral. This chapter proves a stronger result for the mutual link formation model.

This is how to read this chapter. We first characterize the equilibrium networks of one-shot games and then the subgame-perfect equilibria of the repeated games. A number of proofs use lengthy combinatorics, mostly to deal with the existence of networks having specific properties. These proofs are not essential and we gather them in

Appendix A. The proofs that carry important arguments are placed in the main sections. To support the intuition we supplement formal analysis with a sequence of examples. Numerical examples are often escorted also by graphical illustrations. A quick, but not comprehensive, overview of our results can be acquired by reading through descriptions of the models and following our analysis through the examples.

4.2 Network Dilemma Game with Linking Constraints

In the basic game players are free to propose and establish any number of links to any subset of other players. In this chapter we consider the games in which players may be constrained in the number of links they can propose.

Let k_i be the integer denoting the maximal number of links that player i can propose. We refer to $\mathbf{k} = (k_1, ..., k_n)$ as the vector of linking constraints of players $N = \{1, .., n\}$. Proposing and establishing links is costless. A player i can propose a link to any other player in the group and can propose up to k_i links.

Each player also chooses an action in a prisoner's dilemma game. Players cannot discriminate in their actions.[1] One game is played over each established link. For simplicity (and with no loss of generality) we assume that the value of the outside option is $o = 0$. Hence, following classification from Chap. 3 we say that outside option is *low* if $d > 0$, and *high* if $d < 0$.

Let $A_i = \{C, D\}$ be the set of actions and let

$$P_i(k_i) = \{p_i \in \{0,1\}^n \mid p_{ii} = 0 \text{ and } \sum_{j \in N} p_{ij} \leq k_i\}$$

be the set of linking choices of player i. The set of moves of player i is denoted by $J_i(k_i) = A_i \times P_i(k_i)$. Let $J(\mathbf{k}) = \times_{i \in N} J_i(k_i)$ and let $J_{-i}(\mathbf{k}) = \times_{j \in N \setminus \{i\}} J_j(k_j)$ be the set of moves of player i's opponents. When we refer to a game with linking constraints, move refers to a pair $(a_i, p_i) \in J_i(k_i)$ and profile of moves refers to a profile $(a, p) \in J(\mathbf{k})$. For each profile of moves (a, p) the payoff to player i is given by

$$\pi_i(a, p) = \sum_{j \in L_i(p)} v(a_i, a_j) = \sum_{j \in N} v(a_i, a_j) g_{ij}(p) . \tag{4.1}$$

[1] In Sect. 4.5.2 we show that all results of this chapter apply also if players are free to choose different actions in interactions with different neighbors, as long as information about past action choices is complete.

In line with (3.2), we can also define the payoff to player i, given network g and a profile of actions a, by

$$\pi_i(a,g) = \sum_{j \in L_i(g)} v(a_i,a_j) = \sum_{j \in N} v(a_i,a_j) g_{ij} \; . \tag{4.2}$$

We refer to the stage game $\Gamma(\mathbf{k}) = \langle N, J(\mathbf{k}), \pi \rangle$ as the *network dilemma game with linking constraints*, or shortly, the *dilemma game.*

We do not consider the possibility that players may randomize in their choices. Randomization can be analytically useful, e.g. to smooth the response functions, to make the action sets convex and, consequently, to ensure the existence of Nash equilibria. In our view there is no need for this in the analysis of network dilemma games, because at least one Nash equilibrium in pure strategies always exists. Including equilibria in mixed strategies would add little to our results but significantly complicate the notation and analysis. Furthermore, in repeated games the mixed strategies are difficult to enforce because deviations from mixed strategies cannot be detected. This is usually resolved either by assuming that each randomization process itself can be observed or that players condition their moves on the outcome of some public randomization mechanism. However, for the games in this chapter it seems more intuitive to assume that only pure strategies are used.

Move $(a_i^*, p_i^*) \in J_i(\mathbf{k})$ is a *best response of player* i to the profile of moves $(a_{-i}, p_{-i}) \in J_{-i}(\mathbf{k})$ if

$$\pi_i(a_i^*, p_i^*, a_{-i}, p_{-i}) \geq \pi_i(a_i', p_i', a_{-i}, p_{-i})$$

for any $(a_i', p_i') \in J_i(\mathbf{k})$. A *Nash equilibrium* of game $\Gamma(\mathbf{k})$ is a profile of moves $(a^*, p^*) \in J(\mathbf{k})$ such that for each player i,

$$\pi_i(a^*, p^*) \geq \pi_i(a_i, p_i, a_{-i}^*, p_{-i}^*) \tag{4.3}$$

for any $(a_i, p_i) \in J_i(k_i)$. Network g is an *equilibrium network* if there exists a Nash equilibrium (a^*, p^*) such that $g = g(p^*)$.

The game obtained when a network dilemma game with linking constraints $\Gamma(\mathbf{k})$ is repeated T-times is denoted by $\Gamma^T(\mathbf{k}) = \langle N, J(\mathbf{k}), \pi, T \rangle$. Recall that when discussing a repeated game Γ^T the constituent game Γ is the *stage game* and the equilibria of Γ are the *static equilibria.* We refer to the repeated game $\Gamma^T(\mathbf{k})$ as the *finitely repeated dilemma game with linking constraints* or, shortly, the *repeated dilemma game.*

The profile of moves at t is denoted by (a^t, p^t). The action of player i in period t is a_i^t and her linking choice at t is p_i^t. The history at the end of time t is the sequence of moves $h^t = ((a^1, p^1), ..., (a^T, p^T))$. The total payoff to player i at the end of the repeated dilemma game is determined by the terminal history h^T and given by

$$\Pi_i(h^T) = \sum_{t=1}^{T} \pi_i(a^t, p^t) .$$

A strategy for player i in the game $\Gamma^T(\mathbf{k})$ is a function σ_i which selects, for any history of play, an element of $J_i(\mathbf{k})$. We say that a strategy profile is *semi-cooperative* if at least one semi-cooperative relation (see Sect. 3.1) is established in an outcome of at least one period. Otherwise, we say it is *defective.* We say that a strategy profile is *cooperative* if, in an outcome of at least one period, (i) each player has at least one neighbor and (ii) all players cooperate.

The following proposition describes some obvious properties of subgame-perfect equilibria of network dilemma games. It states that in any such equilibrium all players defect in the last period of the game. Therefore, a subgame-perfect equilibrium is cooperative only if there is an early period in which all players cooperate. The cooperative subgame perfect equilibria that we describe in this chapter will be such that (i) all players cooperate during all initial periods, (ii) then sequentially or simultaneously turn to defection, until (iii) all players defect during one or more final periods. As this proposition shows, if such an equilibrium is possible for a network dilemma game repeated T-times, then one exists also when the game is repeated more than T-times.

Proposition 4.1. *If a cooperative subgame perfect equilibrium exists for $\Gamma^{T'}(\mathbf{k})$ then there exists an integer $\gamma \leq T'$ such that a cooperative subgame perfect equilibrium for $\Gamma^T(\mathbf{k})$ exists if and only if $T \geq \gamma$. In any subgame perfect equilibrium, (i) at least one relation is defective in each of the final $\gamma - 1$ periods, and (ii) all relations are defective in the final period.*

Proof. Assume that a cooperative subgame perfect equilibrium (CSPE) exists for $\Gamma^{T'}(\mathbf{k})$. Obviously, there exists a smallest γ such that a CSPE exists for $\Gamma^\gamma(\mathbf{k})$. In any one-shot dilemma game all relations are defective, hence $\gamma \geq 2$. For any T : if a CSPE σ exists for $\Gamma^T(\mathbf{k})$ then one exists also for $\Gamma^{T+1}(\mathbf{k})$; for example, "play σ in the first T periods and play a static equilibrium in period $T + 1$".

A subgame perfect equilibrium (SPE) of $\Gamma^T(\mathbf{k})$ must select a SPE for any subgame $\Gamma^t(\mathbf{k}), t < T$ and any history. No CSPE exist for $\Gamma^t(\mathbf{k})$ if $t < \gamma$. All relations are defective in any Nash equilibrium of $\Gamma(\mathbf{k})$.

Hence, an equilibrium outcome path for the $\Gamma^T(\mathbf{k})$ must be such that all relations are defective in the last period, and in none of the last $\gamma - 1$ periods are all relations cooperative. □

Consider the mutual link formation model. For a profile of linking choices p the network of established links $g(p)$ is defined by $g_{ij}(p) = \min\{p_{ij}, p_{ji}\}$.

For each network g and each player i let $p_i(g)$ be the binary vector defined by $p_{ij}(g) = g_{ij}$ for all j. We say that player i *supports network* g when her linking choice is $p_i(g)$. The profile $p(g) = (p_1(g), ..., p_n(g))$ is the minimal *support* for the network g, in the sense that each link proposal in $p(g)$ is necessary to establish g.

To avoid trivialities we assume that $c > 0$, that is, the value of mutual cooperation is higher than the outside option. We also assume that $1 \leq k_i \leq n - 1$ for each player i.[2] Let

$$\mathcal{G}(\mathbf{k}) = \{g \mid l_i(g) \leq k_i \text{ for each } i\} \tag{4.4}$$

be the set of feasible networks. For reference we restate the Prisoner's dilemma payoff matrix:

	Player j C	D
Player i C	c, c	e, f
D	f, e	d, d

4.3 Equilibria and Linking-Proof Networks

The complete characterization of Nash equilibria for dilemma games is possible but cumbersome. The following proposition characterizes the actions and the networks that are established in a Nash equilibrium of a one-shot dilemma game. The set of equilibrium networks depends only on whether the outside option is low or high.

Proposition 4.2. *Consider the dilemma game* $\Gamma(\mathbf{k})$.

1. *A player cooperates in a Nash equilibrium only if no links to her are proposed. Hence, in a Nash equilibrium all cooperative players are isolated.*

[2] Obviously, in a group with n players each of them can establish at most $n - 1$ links, hence it makes sense to assume that $k_i \leq n-1$ for all i. On the other hand, if a player cannot make any links ($k_i = 0$) she will be isolated regardless of the linking behavior of other players. An isolated player does not affect the dynamics of play. We therefore assume $k_i \geq 1$ for all i.

2. *If* $d \geq 0$ *then* $(\mathbf{D}, p(g))$ *is a Nash equilibrium for any network* $g \in \mathcal{G}(\mathbf{k})$.
3. *If* $d < 0$ *then the empty network is the unique equilibrium network.*

Proof. (1) Let (a^*, p^*) be a Nash equilibrium (NE). There cannot be a pair of players i, j such that $a_i^* = C$ and $p_{ji}^* = 1$ as in this case i could strictly increase her payoff by switching to $a_i = D$ and $p_{ij} = 1$. There also cannot be a pair of players i, j such that $a_i^* = C$ and $p_{ij}^* = 1$ and $p_{ji}^* = 0$ as in this case j could strictly increase her payoff choosing some p_j such that $p_{ji} = 1$.

(2) For any $g \in \mathcal{G}(\mathbf{k})$ the $p(g)$ is the minimal support for g in the sense that all proposed links are reciprocated. Hence, no player can increase her payoff by unilaterally proposing additional links and can only (weakly) decrease her payoff if she removes some links. Obviously, no player can increase her payoff by switching to cooperation. The profile $(\mathbf{D}, p(g))$ is therefore a NE.

(3) According to (1), if in NE players i and j are neighbors they should both defect. However, each could then increase her payoff by removing the link. Hence, in a NE there are no neighbors. □

Statement (1) asserts that in a Nash equilibrium of a dilemma game all relations are defective. That is, in a Nash equilibrium players with at least one established link defect but the isolated players may cooperate. Consequently and as stated in (3), if the mutual defection payoff d is negative no links are established in a Nash equilibrium. For non-negative d, however, statement (2) asserts that any feasible network is an equilibrium network. The dilemma game with $d = 0$ is rather trivial. In the rest of the section we consider only games with $d \neq 0$.

Consider a dilemma game with $d > 0$. If all players defect each of them maximizes her payoff by establishing as many links as possible. It may thus be surprising that networks with relatively little or no links can be established in a Nash equilibrium. The multiplicity of equilibrium networks is a consequence of the mutual link formation assumption. A link between two players is established only if both propose it. If one of the players does not propose the link the other is indifferent between proposing the link or not. It may therefore happen that two players prefer to establish a mutual link but none proposes it because each knows that the other will not propose the link. Hence, in a Nash equilibrium there may exist a pair of players such that (i) each could beneficially establish at least one more link but (ii) none proposes a link to the other player. This appears especially unintuitive given that proposing links is costless (see e.g. the discussion in [83]).

Several resolutions have been discussed in the related literature on network formation [76, 62, 77]. We follow the common approach in this literature and supplement the idea of the Nash equilibrium with a requirement of network stability. Roughly speaking, an equilibrium network is said to be stable if no pair of separated players exists such that, by adding the mutual link, each could strictly increase her payoff.

Our definition of network stability is inspired by the concept of pairwise-stability, introduced by Jackson and Wolinsky [59]. It is defined for network formation games with mutual linking, but with exogenously fixed cost and benefit rules. While it retains some flavor of the non-cooperative equilibrium, pairwise-stability is analyzed with the tools of cooperative game theory.[3] It is not immediately obvious how to appropriately generalize pairwise-stability to network formation games with endogenous benefit structure, and how to analyze such games with the tools of non-cooperative game theory. Our definition of linking-proofness below is chosen because it eliminates the unintuitive Nash equilibria in the dilemma games.

In a dilemma game the value of a link depends on the actions taken by the linked players. The dilemma game in this section is then a network formation game with mutual linking, no linking costs, and an endogenous benefit structure. The players are free to deviate by si-

[3] Below we give a slightly simplified definition of pairwise-stability (for the general definition and discussion see [41]). The network formation game in the cooperative form is characterized by the set of feasible networks $\mathcal{G}$ and by the allocation rule $Y : \mathcal{G} \to \mathbb{R}^n$ which describes how the value generated by the network is shared among the players. The network $g \in \mathcal{G}$ is said to be *pairwise stable* with respect to the allocation rule Y if

(i) for all $ij \in g$, $Y_i(g) \geq Y_i(g \ominus ij)$ and $Y_j(g) \geq Y_j(g \ominus ij)$, and
(ii) for all $ij \notin g$ such that $(g \oplus ij) \in \mathcal{G}$, $Y_i(g \oplus ij) > Y_i(g)$ implies $Y_j(g \oplus ij) < Y_j(g)$.

It is implicit in the definition of pairwise stability that the termination of a link can be done unilaterally but that the addition of a link requires mutual consent of both involved players. A network is pairwise-stable if no player can profit by unilaterally removing a link and no pair of players can simultaneously profitably deviate by adding a mutual link.

A few extensions of the original definition of pairwise-stability exist. For example, one may assume that adding a link is plausible if it makes both linked players strictly better off, or alternatively, if it makes both linked players weakly better off, or yet some intermediate concept, see [84]. The original definition does not consider deviations of one player by simultaneous removal of several links or by simultaneous removal of one link and addition of another link, which would bring the concept closer to the spirit of a non-cooperative equilibrium (see the recent approaches in [76, 62]).

multaneously changing their action, removing links, and adding a new link.

A profile of moves is a Nash equilibrium if no player wants to deviate given the moves of the other players. We say that an equilibrium profile of moves is linking-proof if no player wants to deviate even if there was a possibility to establish a new link. In other words, the only multilateral deviation considered is that of adding a new link. However, a player is free to deviate by adding a new link in parallel to any other unilateral changes to her move.

For a profile of linking choices p let $p \oplus ij$ be another profile defined as

$$(p \oplus ij)_{i'j'} = \begin{cases} 1 & \text{if } i'j' = ij \\ p_{i'j'} & \text{otherwise.} \end{cases}$$

In words, $p \oplus ij$ is a profile of linking choices obtained from p by assuming that player i proposes a link to j in addition to all the links proposed according to p. If $p_{ij} = 1$ then $p \oplus ij = p$.

Definition 4.1. *The profile of moves $(a^*, p^*) \in J(\mathbf{k})$ is a **linking-proof equilibrium (LP equilibrium)** of game $\Gamma(\mathbf{k})$ if (i) for each player i her move (a_i^*, p_i^*) is a best response to (a^*, p^*), and (ii) for each pair of players i, j either (a_i^*, p_i^*) is a best response to $(a^*, p^* \oplus ji)$ or (a_j^*, p_j^*) is a best response to $(a^*, p^* \oplus ij)$. Network $g^* \in \mathcal{G}(\mathbf{k})$ is **linking-proof** if it is established in some linking-proof equilibrium.*

Any linking-proof equilibrium is Nash equilibrium, due to (i). Furthermore, in a linking-proof equilibrium there is no pair of separated players such that each of them strictly prefers to change her move and establish the mutual link. In other words, in each pair of separated players at least one of them has no move that strictly improves her payoff and establishes the mutual link.

Supplementing an equilibrium concept with that of network stability is not the only route to study strategic form games of network formation with mutual linking. For example, a stronger equilibrium concept, such as coalition-proof Nash equilibrium [85] or Strong Nash equilibrium [86], may also eliminate unintuitive Nash equilibria, see [84, 83, 87, 88]. The problem with these equilibrium refinements is that they eliminate too many networks because they require stability against deviations by variously sized coalitions of players. Our concept of linking-proof equilibrium is much weaker: it supplements the Nash equilibrium only with the requirement that a pair of players is linked whenever it is possible and in their mutual interest, but does not permit any other coordinated actions.

We denote the set of linking-proof networks of the game $\Gamma(\mathbf{k})$ by $\mathcal{S}(\Gamma(\mathbf{k}))$. The following proposition asserts that, when $d > 0$, the set $\mathcal{S}(\Gamma(\mathbf{k}))$ contains all networks in which only the linking constraints prohibit any pair of separated players from adding a mutual link.

Proposition 4.3. *Consider dilemma game $\Gamma(\mathbf{k})$.*

1. *Let $d > 0$. Network $g^* \in \mathcal{G}(\mathbf{k})$ is a linking-proof network if and only if there is no pair of separated players i, j such that $l_i(g^*) < k_i$ and $l_j(g^*) < k_j$.*
2. *If $d < 0$ the empty network is the unique linking-proof network.*

Proof. (1.a) Let $d > 0$. Consider $g^* \in \mathcal{S}(\Gamma(\mathbf{k}))$ and let (a^*, p^*) be the LP equilibrium such that $g(p^*) = g^*$. Assume (A1): there is a pair of distinct players i and j such that $l_i(g^*) < k_i$ and $l_j(g^*) < k_j$.

Let $p^*_{ij} = 0$. Define p'_i as follows: (a) if there is j' such that $p^*_{ij'} = 1$ and $g^*_{ij'} = 0$ then let p'_i coincide with p^*_i in all values except for setting $p'_{ij} = 1$ and $p'_{ij'} = 0$ (if i proposed a link which was not reciprocated propose instead the link to j)
(b) if there is no such j' then let p'_i coincide with p^*_i in all values except for setting $p'_{ij} = 1$ (i did not propose the maximal number of links, hence propose also the link to j). Define p'_j in a symmetric manner. As $p^* \in P(\mathbf{k})$ then $p'_i \in P_i(k_i)$ and $p'_j \in P_j(k_j)$.

Because $g^*_{ij} = 0$ it must be that either $p^*_{ij} = 0$ or $p^*_{ji} = 0$ or both. If $p^*_{ji} = 1$ then for player i, (D, p'_i) is a better response to (a^*, p^*) than is (a^*_i, p^*_i), as thus the link ij is established in addition to all other links of player i and this increases the payoff to player i for at least $d > 0$. But this should not be possible in a LP equilibrium, hence $p^*_{ji} = 0$. In this case, however, (D, p'_i) is a better response to $(a^*, p^* \oplus ji)$ than is (a^*_i, p^*_i).

Similarly we can show that $p^*_{ij} = 0$, in which case (D, p'_j) is a better response to $(a^*, p^* \oplus ij)$ than is (a^*_j, p^*_j). This, however, is again not possible in a LP equilibrium. We thus proved that (A1) cannot hold.

(1.b) Let $d > 0$. Consider network $g \in \mathcal{G}(\mathbf{k})$ such that $g_{ij} = 1$ for each pair of distinct players i, j such that $l_i(g) < k_i$ and $l_j(g) < k_j$. We argue that $(\mathbf{D}, p(g))$ must be a LP equilibrium. By Proposition 4.2 it is a Nash equilibrium which proves part (i) of the definition of LP equilibrium. We show next that part (ii) is also fulfilled.

Take a pair of distinct players i and j. If $g_{ij} = 1$ then (ii) follows from (i). Assume therefore $g_{ij} = 0$, implying that either $l_i(g) = k_i$ or $l_j(g) = k_j$. W.l.o.g. let $l_i(g) = k_i$. Player i cannot increase her payoff by removing or by moving some of her links. She also cannot add any new links. Hence (D, p_i) is a best response to $(\mathbf{D}, p \oplus ji)$, and (ii) is satisfied.

(2) Let $d < 0$. By Proposition 4.2 the empty network is the only equilibrium network and thus the only candidate to be linking-proof. Let 0_i be the vector denoting that player i proposes no links, and let $\mathbf{0} = (0_1, ..., 0_n)$ be the corresponding profile. To see that $(\mathbf{D}, \mathbf{0})$ is a LP equilibrium, note that for each i, $(D, 0_i)$ is best response both to $(\mathbf{D}, \mathbf{0})$ and to $(\mathbf{D}, \mathbf{0} \oplus ij)$ for any $j \neq i$. □

Let $d > 0$. If a pair of players in a linking-proof network are separated it must be that at least one of them has established her maximal number of links. The restriction to linking-proof equilibria may considerably reduce the set of networks established by an equilibrium profile and lead to existence problems. However, Proposition 4.4 below states that a linking-proof network always exists. When only the empty network is established in a Nash equilibrium (i.e. when $d < 0$) this network is the unique linking-proof network. Several linking-proof networks may exist when $d > 0$. We demonstrate this in the following examples.

Example 4.1 (no linking constraints). Let $d > 0$ and let $k_i = n - 1$ for all i. In this game the linking is not constrained and players may establish any number of links. This constitutes our basic dilemma game Γ. Hence, if $g_{ij} = 0$ then $l_i < k_i$ and $l_j < k_j$ for any pair of distinct players i and j and g is not linking-proof. The complete network is therefore the unique linking-proof network.

Example 4.2 (uniform linking constraint). Let $d > 0$ and let $n = 4$. Let $\mathbf{k} = (2, 2, 2, 2)$. Any of the 64 possible networks can be established in a Nash equilibrium. However, only seven of them, illustrated in Fig. 4.1, are linking-proof. Networks g^I, g^{II} and g^{III} constitute the three possible wheel networks in which each player links to two neighbors. In networks g^1, g^2, g^3 and g^4 one player is isolated and each of the other three players has two links. For $i = 1, 2, 3, 4$ the network g^i forms when players $N \backslash \{i\}$ form a clique, thus leaving player i without any links.

The following Proposition asserts that every dilemma game has a linking-proof equilibrium, and consequently, a linking-proof network.

Proposition 4.4. *There always exists a linking-proof equilibrium of the game $\Gamma(\mathbf{k})$.*

Proof. Let 0_i be the vector denoting that player i proposes no links, and let $\mathbf{0} = (0_1, ..., 0_n)$ be the corresponding profile. The profile $(\mathbf{D}, \mathbf{0})$ is LP equilibrium for dilemma games with $d \leq 0$. For $d < 0$ this is

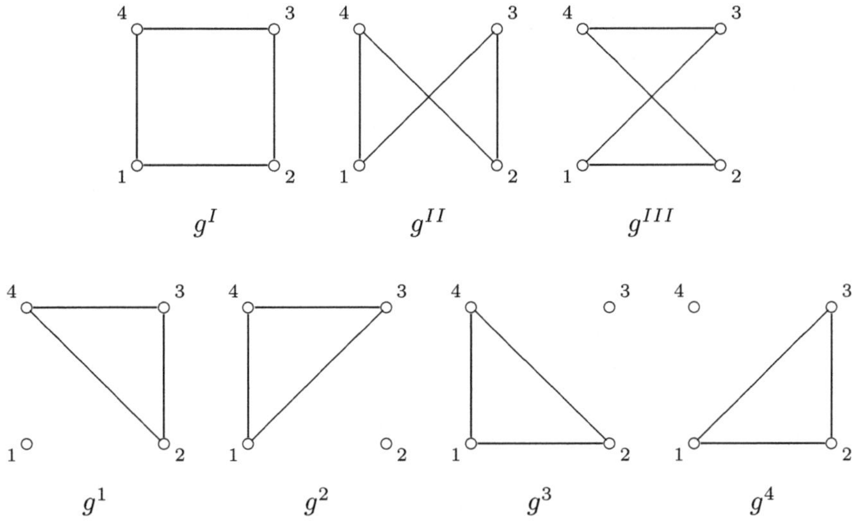

Fig. 4.1. Linking-proof networks of the dilemma game with $n = 4$ players, $\mathbf{k} = (2, 2, 2, 2)$, and $d > 0$. Only established links are shown

shown in Proposition 4.3. For $d = 0$ this is because if all players defect no player can increase her payoff either by switching to cooperation, or by removing, moving or adding links.

For $d > 0$ a LP network can be constructed with the following iterative procedure. Let g^0 be an empty network. For $x = 1, 2, ...$ let g^x be the network obtained from g^{x-1} by the addition of one link: take a pair of distinct players i, j such that $g_{ij}^{x-1} = 0$, $l_i(g^{x-1}) < k_i$ and $l_j(g^{x-1}) < k_j$, and let $g_{ij}^x = 1$, and otherwise let $g_{i'j'}^x = g_{i'j'}^{x-1}$. Stop the procedure if there is no such pair of players.

This procedure stops after finite number of steps, because a new link is added in each step and there is a finite number of possible links. It follows from Proposition 4.3 that the resulting network is an LP equilibrium. □

4.3.1 Subgame Perfect Equilibria of the Repeated Network Dilemma Game

In the repeated game the players can condition their linking choices on the behavior of players in previous periods. More specifically, players may punish a deviation from the agreed outcome path by removing links to any player that deviated. For example, players may agree to cooperate and punish defectors by exclusion from the neighborhood of cooperative players.

In a repeated game we may define exclusion as follows. Let $t \geq 2$. We say that player i *excludes* player j in period t if i and j were neighbors at $t-1$ and i removes the link with j in period t.

Two players may exclude each other simultaneously. For example, player i may remove her link with player j if she knows j will also remove it. Recall that each player is indifferent between proposing a link or not if she knows that it will not be reciprocated. Two neighbors could therefore exclude each other even if the resulting network is such that each would prefer to keep the link established. We call an exclusion *mutual* if two players exclude each other but both miss at least one link in the resulting network.[4]

On the other hand, a single player may be willing to exclude another regardless of the other player's linking decision. It is possible, for example, to exclude a player and yet establish the maximal number of links. This can be done, for example, by removing the link with one player and establish a new one with another player, thus excluding the first and including the latter.[5] We say that player i *unilaterally* excludes player j if i still establishes her maximal number of links in the resulting network. We call an exclusion *unilateral* if each *excluding* player establishes her maximal number of links in the ensuing network. There is no condition on the number of links that the excluded player establishes in the ensuing network. Unilateral exclusion is defined only for games with mutual link formation and is not related to unilateral link formation. See Fig. 4.2 for examples of unilateral and mutual exclusion.

We begin with a few examples of the equilibria of repeated dilemma games. The first example illustrates that not every Nash equilibrium of the repeated dilemma game is subgame perfect. Recall that in order to distinguish between the Nash equilibria of the repeated game and those of the stage game we refer to the latter as the static equilibria.

Example 4.3. Consider the dilemma game $\Gamma = \langle \{1,2\}, J, \pi \rangle$ with two players and no linking constraints,[6] repeated twice. Let the outside option be low, $d > 0$. Consider the following strategy σ_i for player $i = 1, 2$: "Propose the link and cooperate in the first period. In the second period defect, and propose the link if and only if the opponent cooperated in the first period."

---

4 We say that player i is missing one or more links in network g if she establishes less than her maximal number of links, that is, if $l_i(g) < k_i$.

5 If a player i removes a link with player j to establish a link with player j' we say that i *relocates* a link.

6 This game is a variant of the usual two-player prisoner's dilemma game with outside option.

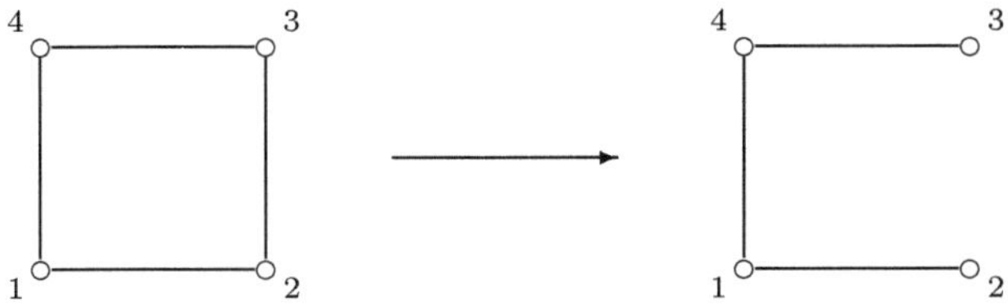

(a) mutual exclusion

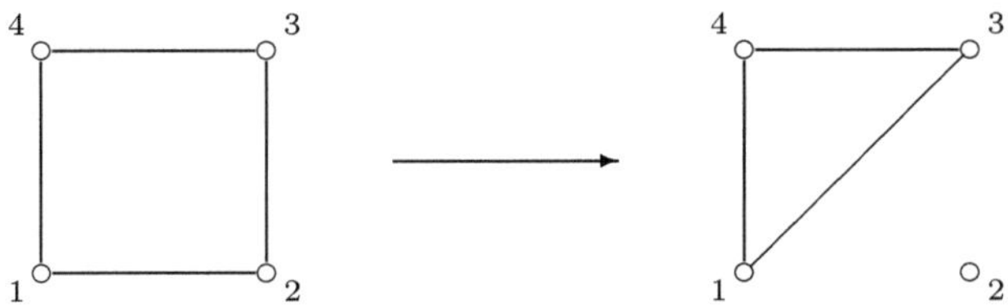

(b) unilateral exclusion

Fig. 4.2. Two types of exclusion under mutual link formation. Exclusion in the repeated dilemma game with $n = 4$ players, $\mathbf{k} = (2, 2, 2, 2)$, and $d > 0$: (**a**) players 2 and 3 simultaneously exclude each other, each missing one link in the resulting network; (**b**) players 1 and 3 both exclude player 2 and establish a mutual link, thus missing no links in the resulting network

The pair (σ_1, σ_2) constitutes a Nash equilibrium of Γ^2 if $f < c + d$: given that player i follows σ_i, the best response of player j entails always proposing the link and defection in the last period; this given, defection in the first period earns her a total payoff of f, which is less than $c + d$ earned if she cooperates in the first period.

Assume now that player 2 defects in the first period nevertheless. In response, the player's strategies require both players to defect, and that only player 2 proposes the link which, consequently, is not established. However, in the absence of commitment it is not obvious that in the second period player 1 will persist in playing σ_1. In fact, rather than to exercise the threat of removing the link, player 1 does better in the second period by reneging on her threat and propose the link anyway. If player 1 is rational she never excludes player 2. If player 2 is rational she anticipates this and defects in both periods. Realizing this, rational players would not even agree on playing (σ_1, σ_2). The threat of unilateral exclusion, implicit in the strategy σ_1, is in this sense empty and the equilibrium (σ_1, σ_2) is not subgame perfect.

The above example illustrates the general discussion in Sect. 2.2.2 and motivates our focus on subgame perfect equilibria. For many dilemma games only trivial subgame-perfect equilibria exist, in the sense that a static equilibrium must be selected in each period. Recall assertion 2 of Lemma 2.1: if a stage game has a unique Nash equilibrium, the finitely repeated game has a unique subgame perfect equilibrium.

Example 4.4 (Exogenous network). The n-player prisoner's dilemma game can be seen as a dilemma game without the linking choice. Rather, the network is fixed exogenously to be a complete network. Each player chooses an action and must play the game with all other players. The payoffs are given by (4.2), with g being the complete network.

For each player defection strictly dominates cooperation. The stage game thus has a unique Nash equilibrium in which all players defect. Consequently, the finitely repeated n-player prisoner's dilemma game has a unique subgame perfect equilibrium in which all players defect in each of the periods.

Assertion 3 of Lemma 2.1 states that if static equilibria are payoff-equivalent the set of the subgame perfect equilibria is characterized by all possible sequences of static equilibrium outcomes. Consequently, a player earns the same total payoff in each of the subgame perfect equilibria.

Example 4.5 (Endogenous network, high outside option). Consider any dilemma game and let $d < 0$. Proposition 4.2 states that the empty network is established in any static Nash equilibrium of this game. All its static equilibria are therefore payoff-equivalent because the payoff of each player in the empty network is 0, regardless of the profile of actions. By implication, each subgame perfect equilibrium of the repeated dilemma game selects a sequence of static equilibria. To summarize, whenever $d < 0$ the empty network is established along the outcome path of any subgame perfect equilibrium of the repeated dilemma game.

All the subgame-perfect equilibria that we constructed above have a common feature: they select a static equilibrium in each period. Statement (1) of Lemma 2.1 asserts that any profile of strategies that in each period, and regardless of the history, selects some static equilibrium is subgame perfect. For any repeated dilemma game a number of subgame perfect equilibria can therefore be constructed in this way, from sequences of static equilibria. However, in each static equilibrium all relations are defective. Hence, all subgame perfect equilibria constructed from sequences of static equilibria are defective.

No other subgame perfect equilibrium exist for finitely repeated dilemma games with high outside option, which is shown in the example above. In the following section we show that cooperative subgame-perfect equilibria may exist for dilemma games with low outside option.

4.4 Cooperation in a Subgame Perfect Equilibrium

In Sect. 4.4.1 we show that, for any repeated dilemma game with $d > 0$, cooperative relations may be sustained in a subgame perfect equilibrium via the threat of mutual exclusion. Mutual exclusion requires that both players simultaneously remove the link with each other. Punishment with mutual exclusion thus requires the participation of a punished player in her own punishment. Furthermore, the resulting network is not linking-proof. This may motivate the players, especially the punished one, to renege on their participation in the punishment. In Sect. 4.4.2 we follow up by discussing the possibility to sustain cooperation via threats of unilateral exclusion. We show that this is possible for most vectors of linking constraints.

4.4.1 Cooperation Through Threats of Mutual Exclusion

Consider a dilemma game $\Gamma(\mathbf{k})$ and let there be a network $g^* \in \mathcal{G}(\mathbf{k})$ with no isolated players: $l_i(g^*) \geq 1$ for each player i. For each i let $g^i = g \ominus 1i \ominus ... \ominus ni$ be the network obtained from g^* by removing all established links with player i: $l_i(g^i) = 0$. Obviously, $g^i \in \mathcal{G}(\mathbf{k})$ for each i. Note that q^i is a Nash equilibrium only if (i) player i proposes no links and (ii) no player proposes a link to i.

Let $d > 0$. It follows from Proposition 4.2 that g^* and all g^i can be established in the Nash equilibria $q^* = (\mathbf{D}, p(g^*))$ and $q^i = (\mathbf{D}, p(g^i))$. The payoff of player i in each of these equilibria is proportional to the number of her neighbors. In particular,

$$\pi_i(q^*) = d\ l_i(g^*) > 0 \text{ and } \pi_i(q^i) = d\ l_i(g^i) = 0\ .$$

Consider now the profile of moves $r^* = (\mathbf{C}, p(g^*))$ where all players cooperate and establish the network g^*. The payoff of player i in this profile is

$$\pi_i(r^*) = c\ l_i(g^*) > 0 = \pi_i(q^i)\ .$$

Recall Definition 2.5, of a trigger strategy profile. Take two integers $T > t^* > 0$. Consider the trigger strategy profile $\sigma(r^*, q^*, \{q^i\}_{i \in N}, T, t^*)$, which can be described in words as follows:

"Cooperate in the early periods $1, ..., T-t^*$ and defect in the remaining periods. Establish network g^* in each period. If in an early period some player i defects, establish network g^i and defect forever."

Each player i earns strictly more in the equilibrium q^* compared to equilibrium q^i. If player i defects early, she earns $t^*\pi_i(q^i) = 0$ in the last t^* periods. If there are no early defections she earns $t^*\pi_i(q^*)$. Hence, if she defects early she looses $t^*\pi_i(q^*)$ in the last t^* periods. There exists a sufficiently large t^* such that for each player this loss is larger than the benefit from any early deviation. If t^* is such, then no player wants to defect in an early period and cooperation is sustained.

Proposition 4.5 confirms this conclusion: there exist T, t^*, such that $\sigma(r^*, q^*, \{q^i\}_{i\in N}, T, t^*)$ is a subgame perfect equilibrium of $\Gamma^T(\mathbf{k})$. By definition, $\sigma(r^*, q^*, \{q^i\}_{i\in N}, T, t^*)$ is cooperative as it induces the cooperative outcome r^* in the early periods of the repeated game.

In the following we focus on the trigger strategy profiles where, similar to the profile outlined above, an early defection triggers a change of the network. For reference we use the notation

$$(a, g) = (a, p(g))$$

and define, for networks $g, g^*, g^1, ..., g^n \in \mathcal{G}(\mathbf{k})$, the following trigger strategy profile

$$\varrho(g, g^*, \{g^i\}_{i\in N}, T, t^*) \equiv \sigma((\mathbf{C}, g), (\mathbf{D}, g^*), \{(\mathbf{D}, g^i)\}_{i\in N}, T, t^*) \,. \quad (4.5)$$

Proposition 4.5. *Consider the dilemma game $\Gamma^T(\mathbf{k})$ with $d > 0$ such that there exists a network $g^* \in \mathcal{G}(\mathbf{k})$ with no isolated players. For each i let g^i be the network obtained from g^* by removing all links with player i. For large enough T there exists t^* such that the trigger strategy profile $\varrho(g^*, g^*, \{g^i\}_{i\in N}, T, t^*)$ for $\Gamma^T(\mathbf{k})$ is fully cooperative and subgame perfect.*

Proof. We have outlined the proof in the discussion above. Profiles of moves $(\mathbf{D}, p(g^*))$ and $(\mathbf{D}, p(g^i))$ are Nash equilibria. Furthermore, $\pi_i(\mathbf{C}, p(g^*)) > \pi_i(\mathbf{D}, p(g^*)) > \pi_i(\mathbf{D}, p(g^i))$. The rest of the proof follows from Theorem 2.1 in Chap. 2. □

Corollary 4.1. *A cooperative trigger strategy profile for $\Gamma^T(\mathbf{k})$ with $d > 0$ exists for sufficiently large T if $k_i \geq 1$ for each i and, if n is odd, $k_j \geq 2$ for some j.*

Proof. If n is even, a network $g^* \in \mathcal{G}(\mathbf{k})$ with no isolated players may be established by linking pairs of players: for example, for each $x =$

$1, .., n/2$, players with indices $2x - 1$ and $2x$ are linked. If n is odd a network $g^* \in \mathcal{G}(\mathbf{k})$ with no isolated players may be established by linking two players to player j and link pairs of the remaining players: for example, if $k_1 \geq 2$ players 1 and 2, players 1 and 3, and for each $x = 2, .., n/2$, players with indices $2x$ and $2x + 1$ are linked. □

The trigger strategies of Proposition 4.5 threaten to punish deviations in the early periods with mutual exclusion: after a player deviates from the equilibrium path all links with that players are mutually removed and she remains (mutually) isolated from the network throughout the remaining periods of the game.[7]

Example 4.6 (cooperation via mutual exclusion). Consider again the dilemma game $\Gamma = \langle\{1, 2\}, J, \pi\rangle$ of Example 4.3, repeated twice. Let $f < c + d$ so that, for each player, mutual cooperation in the first and defection in the second period is better than defection in the first and exclusion in the second period. Inspired by Proposition 4.5 we consider the following strategy: "Propose the link and cooperate in the first period. In the second period defect, and propose the link if and only if both players cooperated in the first period."

This strategy differs from the strategy σ_i of Example 4.3 in that it requires second period link removal after deviation of any player, including own deviation. A deviation in the first period is punished by mutual, rather than unilateral exclusion of strategy σ_i. The threat of exclusion is effective. Mutual exclusion is credible as no player can profit by unilateral abstention from excluding the other player. A pair of strategies described above is thus subgame perfect. It is important to note that punishment by mutual exclusion requires participation of both players.

The finitely repeated prisoner's dilemma game without an outside option has a unique subgame perfect equilibrium in which both players defect in all periods. Example 4.6 demonstrates that the addition of a low outside option to the prisoner's dilemma game is sufficient to obtain cooperative subgame perfect equilibria. Cooperation in these equilibria is sustained by the threat of mutual exclusion.

---

[7] Not all deviations need to be punished actively. Some deviations may decrease the payoff of the deviating player and need no subsequent punishment to be deterred. For example, if the play reaches a static equilibrium there is no deviation that increases the payoff of the deviating player. In this case proceeding, after some deviation, with the play as if no deviation occurred already constitutes an effective threat.

Mutual exclusion may not be the most intuitive form of punishment, because it requires participation of a punished player in her own punishment. Following her deviation the player is expected to remove all of her links and isolate herself throughout the remaining periods. Her isolation is an equilibrium because other players never again propose the link to her so that she is indifferent between proposing any links or not. However, there are several reasons why it may not be sensible to expect a player to contribute to her own punishment.

A player may be expected to participate in her own punishment only if it is in her own interest to do so, that is, if doing so constitutes the subgame perfect equilibrium across the remaining periods of the game. To make a threat credible it is sufficient that each player *weakly* prefers to participate in the punishment. This is the case with the threats in the trigger strategies of Proposition 4.5. Namely, knowing that she is being excluded by all other players the punished player is indifferent between participating in the punishment or not, hence she does participate in equilibrium. Similarly, knowing that the punished player will participate in the punishment all other players are indifferent between punishing or not, hence they punish in equilibrium.

However, the punished player never strictly prefers to remove a link in order to support her punishment. She will keep her links if even a smallest chance exists that the other players will not exclude her. Given that after the deviation all players defect, each player strictly prefers to have more links from having less. Proposing a maximal number of links thus never hurts. Yet, punishments with mutual exclusion, as described in Proposition 4.5, always requires that punishing players propose less than their maximal number of links and that the punished player proposes no links. If there is even a slight doubt about whether one of the parties will participate in the punishment, or about whether mutual participation is common knowledge, none of the parties will punish.

In the following section we discuss strategies with threats of unilateral exclusion. Such threats may not always be credible, as we demonstrated in Example 4.3. Nevertheless, we conclude that for almost all dilemma games with linking constraints it is possible to sustain cooperation in the subgame perfect equilibrium with threats of unilateral exclusion only. In such equilibria the punishing players are willing to exclude the punished player regardless of her linking behavior.

4.4.2 Cooperation Through Threats of Unilateral Exclusion

The ideas in this section may be outlined as follows. Consider an n-player dilemma game $\Gamma(\mathbf{k})$ and let $\mathbf{k}$ be such that linking is strictly constrained, that is, $k_i < n-1$ for each player i. Each player must thus select at least one other player to whom she does *not* propose a link. If the dilemma game is repeated the choice of which player to exclude may be made contingent on player's past behavior. For example, players may unilaterally exclude those other players who defected in the previous period. This in turn may be sufficient to discourage early defections.

The following example demonstrates that unilateral exclusion can indeed form a credible threat when linking is constrained. The example also gives the sketch of the general analysis.

Example 4.7 (cooperation via unilateral exclusion). Consider the dilemma game of example 4.2 with $d > 0$, $n = 4$ and $\mathbf{k} = (2,2,2,2)$ for each i. Let the game be repeated twice and let $f < c + d$. Consider the following strategy for player i: "Cooperate in the first and defect in the second period. Establish the network g^I in the first period. In the second period: establish g^I if in the first period (i) no player defected, or (ii) two or more players defected; establish g^j if another player $j \neq i$ was the only player who has defected; make any linking choice otherwise (i.e. if i was the only player to have defected)."

If all players follow this strategy the wheel network g^I forms in both periods, players cooperate in the first period and defect in the second period. If in the first period player i is the only to defect, she is unilaterally excluded by her neighbors and network g^i is established in the second period. See Fig. 4.3 for an illustration.

Network g^i is linking-proof. The opponents of player i are willing to establish g^i independently of i's linking choice. The threat of unilateral exclusion is both credible and effective and the ensuing strategy profile is subgame perfect.

In the example above all threats with exclusion implement unilateral exclusion. A linking-proof network is established in each period along the equilibrium path, as well as in each period along any punishment path. The following proposition asserts that sequential formation of linking-proof networks implies that any exclusion is unilateral.

Proposition 4.6. *Consider a repeated dilemma game $\Gamma^T(\mathbf{k})$ with $d > 0$, and an arbitrary $t \leq T$. If all networks which form along the history h^t are linking-proof, then each exclusion during h^t was unilateral.*

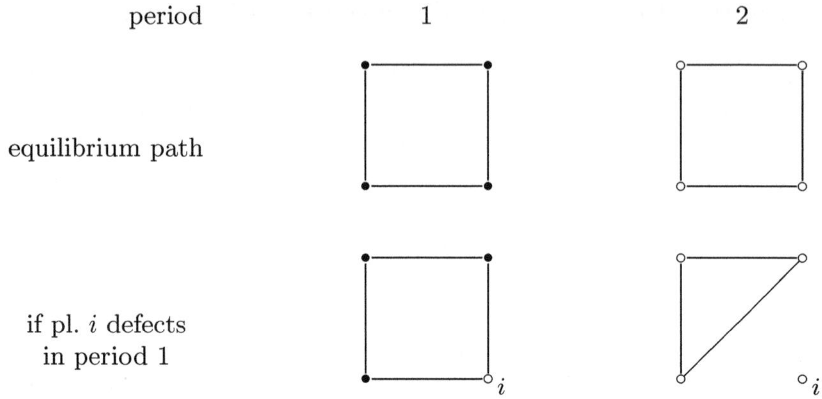

Fig. 4.3. An illustration of a subgame perfect, cooperative and linking-proof strategy profile for the repeated game of example 4.7, with $n = 4$, $\mathbf{k} = (2, 2, 2, 2)$, $d > 0$ and $f < c + d$. Players cooperate (*dark circle*) in the first period and defect (*empty circle*) in the second period

Proof. Assume that, for some $\tau \leq t$, player i excludes player j between periods $\tau - 1$ and τ but none of them establish their maximal number of links in the ensuing network g^τ. Then, $l_i(g^\tau) < k_i$ and $l_j(g^\tau) < k_j$, while $g^\tau_{ij} = 0$. Following Proposition 4.3 g^τ is not a LP network. Hence, g^τ is a LP network only if all exclusions between periods $\tau - 1$ and τ are unilateral. □

To assure that each exclusion is unilateral we thus concentrate on strategies such that, after any possible history, a linking-proof network is established in each period along the resulting outcome path.[8]

Definition 4.2. *Let σ be a profile of strategies in the repeated dilemma game with linking constraints $\Gamma^T(\mathbf{k})$. Profile σ is a* ***linking-proof strategy profile*** *if, for each period $t < T$ and each possible history $h^t \in H^t$, $g(p(\sigma^{t+1}(h^t)))$ is a linking-proof network.*

In what follows we restrict our interest to a subset of subgame perfect equilibria, by focusing on linking-proof subgame perfect profiles. Example 4.7 demonstrates that simple linking-proof subgame perfect equilibria can be constructed using trigger strategies. When trigger strategies fail to sustain cooperation, non-trigger strategies may be used.

[8] We do not assume that along the outcome paths a linking-proof outcome is chosen but only that a linking-proof network is established. Obviously, whenever a pairwise equilibrium is chosen all relations are defective.

We first characterize in Propositions 4.7 and 4.8 the games for which a linking-proof subgame perfect equilibrium can be constructed using simple trigger strategies. We then show in Theorem 4.1 that cooperative and linking-proof subgame perfect equilibria can be constructed for most dilemma games with $d > 0$, using *recursive trigger* strategies.

Trigger strategies

Consider a trigger strategy profile $\varrho(g^*, g^*, \{g^i\}_{i \in N}, T, t^*)$, defined by (4.5). Along the equilibrium path of $\varrho(g^*, g^*, \{g^i\}_{i \in N}, T, t^*)$ network g^* is established in each period, all players cooperate during the early periods $1, ..., T - t^*$ and defect during the remaining periods. Any deviation during the early periods triggers a change of the network: during the remaining periods all players defect and establish network g^i, where i is one of the players which deviated. Profile $\varrho(g^*, g^*, \{g^i\}_{i \in N}, T, t^*)$ is linking-proof whenever each of the networks $g^*, g^1, ..., g^n$ is linking-proof. In this case all threats are credible because in each period of each threat the players play a linking-proof equilibrium.

The following proposition states that a subgame-perfect and linking-proof trigger strategy profile exists if and only if there is a linking-proof network from which each player can be unilaterally excluded.

Proposition 4.7. *Consider an n-player dilemma game $\Gamma(\mathbf{k})$ with $d > 0$. Let $g^*, g^1, ..., g^n$ be linking-proof networks. There exist T, t^*, such that $\varrho(g^*, g^*, \{g^i\}_{i \in N}, T, t^*)$ is subgame-perfect for $\Gamma^T(\mathbf{k})$, if and only if for each i,*

$$l_i(g^i) < l_i(g^*) \ .$$

In this case, for any $T' \geq T$, $\varrho(g^*, g^*, \{g^i\}_{i \in N}, T', t^*)$ is subgame perfect for $\Gamma^{T'}(\mathbf{k})$.

Proof. Let $d > 0$. Assume that $l_i(g^*) > l_i(g^i)$ for some i. If player i defects in one of the early periods she thereby looses some of her neighbors. This decreases all her future per-period earnings,

$$\begin{aligned} \pi_i(\mathbf{D}, p(g^i)) &= d\ l_i(g^i) < d\ l_i(g^*) = \pi_i(\mathbf{D}, p(g^*)) \\ &< c\ l_i(g^*) = \pi_i(\mathbf{C}, p(g^*)), \end{aligned} \tag{4.6}$$

which makes this threat of unilateral exclusion effective for t^* sufficiently small. If $l_i(g^*) > l_i(g^i)$ holds for each player i then (4.6) holds for each i. The profile $\varrho(g^*, g^*, \{g^i\}_{i \in N}, T, t^*)$ then satisfies conditions of Theorem 2.1 to be subgame-perfect for all sufficiently high T and t^* such that $T > t^*$.

Assume now that $l_i(g^*) \leq l_i(g^i)$ for some i. Following $\varrho(g^*, g^*, \{g^i\}_{i \in N}, T, t^*)$ all players cooperate during periods $1, ..., T - t^*$ and defect in the remaining periods. In this case player i earns $l_i(g^*)((T - t^*)c + t^*d)$. If, however, player i defects already in period $T - t^*$, and follows ϱ otherwise, she earns

$$
\begin{aligned}
l_i(g^*)((T - t^* - 1)c + f) + l_i(g^i)t^*d &> l_i(g^*)((T - t^*)c) + l_i(g^i)t^*d \\
&\geq l_i(g^*)((T - t^*)c) + l_i(g^*)t^*d
\end{aligned}
$$

thus strictly increasing her earning. The threat against her defection in period $T - t^*$ is not effective and $\varrho(g^*, g^*, \{g^i\}_{i \in N}, T, t^*)$ is not subgame-perfect. □

When $d > 0$ the vector of linking constraints $\mathbf{k}$ determines the set of linking-proof networks $\mathcal{S}(\Gamma(\mathbf{k}))$. It would thus be of interest to know which vectors $\mathbf{k}$ imply the conditions of Proposition 4.7. Providing a full characterization appears to be a difficult combinatorial task and is beyond the scope of this chapter. Instead, we give below a few partial characterizations and an interpretation of our results.

Proposition 4.8. *Let $d > 0$. Given a sufficient number of periods, trigger strategies can be used to construct (i) cooperative, (ii) linking-proof, and (iii) subgame perfect equilibrium for a repeated dilemma game with linking constraints $\mathbf{k}$, in any of the following cases:*

1. *$k_i = k$ for each i, where $2 \leq k \leq n - 2$,*
2. *$2 \leq k_i < \sqrt{2n - 9}$ for each i.*

In words, cooperation and linking proofness can be sustained (1) if linking constraints are uniform, or (2) if the set of all players is large relative to the maximal possible number of individual links. The proof of Proposition 4.8 follows a sequence of lemmas and is given in Appendix A.

The conditions derived in Proposition 4.8 exclude some interesting dilemma games. Consider, for example, the original network dilemma game with unconstrained linking, $k = n - 1$. In this game the complete network is the unique linking-proof network, as we demonstrated in Example 4.1. Using backwards induction we prove below that the unique subgame-perfect selects this equilibrium in each period. This implies that the complete network is established in each period and all players always defect.

All linking-proof subgame perfect equilibria are defective also when $k_i = 1$ for each i and n is even, because all linking-proof equilibria are payoff-equivalent. In contrast, if $k_i = 1$ for each player i but n is odd, linking-proof subgame perfect equilibrium can be constructed in which all relations are cooperative in early periods.

Proposition 4.9. *Consider an n-player dilemma game $\Gamma(\mathbf{k})$ with $d > 0$ and $k_i = k$ for each i.*

1. *If $k = n - 1$, or if $k = 1$ and n is even, then each linking-proof subgame perfect equilibrium is defective.*
2. *If $k = 1$ and n is odd, there is a linking-proof subgame perfect equilibrium of $\Gamma^T(\mathbf{k})$ with sufficiently large T, such that one player is isolated and all players cooperate in the early periods.*

Proof. (1) Consider the dilemma game without the linking constraints, that is, $k_i = n-1$ for each player i. A complete network g^c is the unique linking-proof equilibrium of this game, and the profile $(\mathbf{D}, p(g^c))$ is the unique static equilibrium establishing g^c. Any linking-proof and subgame perfect profile thus selects $(\mathbf{D}, p(g^c))$ in the last period regardless of the history. By backwards induction, there is a unique such profile and it selects $(\mathbf{D}, p(g^c))$ in each period.

A similar conclusion holds for the dilemma game $\Gamma(\mathbf{k})$ with an even number of players n and $k_i = 1$ for each player i. Each linking-proof network consists of isolated pairs, where each player is linked to one other player. In a linking-proof equilibrium all players defect. This implies that each player earns d in each linking-proof equilibrium profile. All pairwise-equilibria are thus payoff-equivalent. Any linking-proof and subgame perfect profile selects a linking-proof equilibrium in the last period. However, as these are payoff-equivalent, the penultimate period is treated just as the last one and a linking-proof equilibrium is again selected. Hence, by backwards induction a static linking-proof equilibrium must be selected in each period along the outcome path of any such strategy profile. Players defect in all periods along the equilibrium path of any linking-proof and subgame perfect profile.

(2) Construct a linking-proof network g as follows: link each odd i to $i + 1$ and leave n unlinked. The profile of the following strategies is linking-proof: "Establish g as long as there have been no deviations or if more than one player simultaneously deviated. Cooperate in periods $1, ..., T - \gamma$ and defect in the remaining periods. If own neighbor is the only player to deviate then unilaterally exclude her and link to the previously isolated player. If isolated, propose link to any player whose neighbor defected."

Take a player i with a neighbor and consider any period $\tau \leq T - \gamma$. Assume that all players $N \backslash \{i\}$ cooperate. If player i defects, she gains

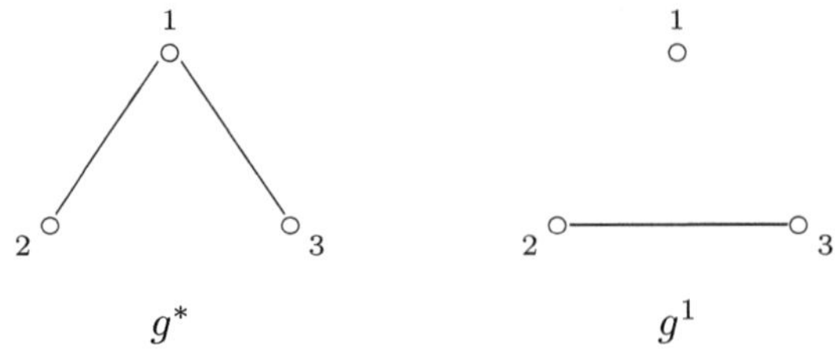

Fig. 4.4. Linking-proof networks of the dilemma game with $n = 3$, $\mathbf{k} = (2, 1, 1)$, and $d > 0$

f immediately and 0 ever after. In contrast, if she cooperates she earns at least $c+\gamma d$. Punishment with exclusion is effective if $f < c+\gamma d$. The strategy profile is subgame perfect and cooperative if $T > \gamma > (f-c)/d$. □

Non-trigger strategies

The trigger strategies above threaten to punish any early defection with unilateral exclusion. If, however, some player cannot be unilaterally excluded the trigger strategies fail to achieve complete cooperation. Take, for example, a player who has the same number of neighbors in all possible linking-proof networks. This player must have the same number of neighbors in each period along the outcome path of any linking-proof profile and cannot be made to cooperate using the trigger strategies defined by (4.5).

Example 4.8. Consider the dilemma game $\Gamma(2, 1, 1)$ with $n = 3$ players and vector of linking constraints $\mathbf{k} = (2, 1, 1)$. Let $d > 0$. There exist only two linking-proof networks, g^* and g^1, illustrated in Fig. 4.4. Players 2 and 3 establish one link in each of these networks and earn the same per-period payoff in the two corresponding stable-equilibria. Since $l_2(g) = 1$ for each linking-proof network $g \in \{g^*, g^1\} \equiv \mathcal{S}(\Gamma(2, 1, 1))$, conditions of Proposition 4.7 are not satisfied. Only player 1 can be punished effectively with unilateral exclusion. Therefore, there exists no cooperative linking-proof equilibrium trigger strategy (4.5) that is subgame-perfect.

It is interesting that it may nevertheless be possible to design strategy profiles that are linking-proof, subgame perfect, and cooperative, albeit not of the form (4.5). We describe the construction of such strategies below.

Trigger strategies threaten to retaliate any deviations with the repetition of a punishment static equilibrium. While such threats are most intuitive, and thus implemented in simple behavioral strategies of the Tit-for-tat or Grim type (see e.g. Chap. 6), they may not be the most effective ones. Trigger strategies can not sustain outcomes which yield payoffs that are below the worst equilibrium payoffs. In contrast, Benoit and Krishna [33] demonstrated that, if the so-called three phase punishment paths are used, any outcome that yields payoffs above the minimax levels may be sustained in early periods of a subgame perfect equilibrium. They proved a limit folk theorem for subgame perfect equilibria of finitely repeated games, under the sufficient condition that each player has several distinct equilibrium payoffs and that the dimension of the set of feasible payoffs is n.[9]

The game is said to have *distinct equilibrium payoffs* if every player has two or more different equilibrium payoffs. This implies that every player has a strictly preferred equilibrium and can be punished by playing her non-preferred one. Conversely, if only one player has distinct equilibrium payoffs then only she can be punished in the last period of the repeated game. Nevertheless, other players may still be induced to play any of their actions, not by threats of punishment but by promises of *rewards.*

An informal outline of the idea can be given as follows. Consider a generic game $\Gamma = \langle N, A, \pi \rangle$ in which only player i has distinct equilibrium payoffs. Because she has distinct equilibrium payoffs player i can be made to play repeatedly any action: only if she complies her preferred equilibrium is played in the last periods. Let $a_i, b_i \in A_i$ be two possible actions of player i and let Γ_{a_i} and Γ_{b_i} be games obtained from Γ by fixing the action of player i to either a_i or b_i. Let j be a player who prefers if player i chooses a_i rather than b_i, in the sense that j earns more in his worst equilibrium of Γ_{a_i} than in his worst equilibrium of Γ_{b_i}. Player j may now be induced to play any of her actions if doing so is rewarded by player i playing a_i. Player j may thus also be made to play any action repeatedly. Hence, both players may be induced to play any of their actions, given a sufficiently large number of periods. A third player, whose payoff crucially depends on the actions of player j, may now in the same fashion be induced to play any of her actions, expecting reward from j. Indeed, iterative rewarding may

[9] The folk theorem was initially proven for infinitely repeated games. It states that any feasible and individually rational payoff vector can be obtained by a long-run undiscounted average of some subgame perfect equilibrium. See [13] and [73] for its early formalizations.

induce any feasible outcome in the early periods of the repeated game if the utilities of *all* players are appropriately interrelated.

See [79] for precise definitions and a thorough elaboration of this idea. Formally, the game is said to have *recursively distinct equilibrium payoffs* if there exists an ordering of the players $1, ..., n$ such that

(i) player 1 has at least two distinct equilibrium payoffs, and
(ii) for all $i < n$, there exist strategy profiles $a(i), b(i) \in A$ such that
  - each player $i+1, ..., n$ plays her best response in both $a(i)$ and $b(i)$,[10] and
  - the payoff to player $i+1$ in $a(i)$ is distinct from her payoff in $b(i)$.

Smith [79] shows that, if the stage game has recursively distinct equilibrium payoffs, any feasible and individually rational outcome may be approximated as an average payoff of the repeated game with sufficiently long horizon. Recursively distinct equilibrium payoffs are thus the necessary and sufficient condition for the limit folk theorem.

We cannot straightforwardly apply the limit folk theorem of Smith to dilemma games because of our additional restriction of linking-proofness. For instance, the strategy profiles designed in [79] punish an early deviation by selecting the outcome in which the deviating player earns her minimax payoff. In the dilemma game with low outside option the player attains her minimax payoff when she has no neighbors. A network in which player has no links may not be linking-proof, though. Indeed, there exist linking constraints such that some players establish a positive number of links in each linking-proof network.[11]

Nevertheless, iterative rewarding can be used to design strategy profiles which are cooperative, subgame-perfect and linking-proof, in dilemma games with linking constraints that do not satisfy the conditions of Proposition 4.7. This is possible under two, albeit weak, conditions: there should exist a player i and two linking-proof networks g^* and g^i such that (i) player i has more neighbors in g^* than in g^i, and (ii) g^* is connected.

The parallel between these conditions and that of the recursively distinct equilibrium payoffs is apparent: condition (i) implies that player i has distinct linking-proof equilibrium payoffs, while condition (ii) implies that payoffs of all players in some linking-proof network are interrelated. The proof of Theorem 4.1, below, is inspired by this parallel. To give the flavor of the proof we construct in the following example

[10] In other words, for each player $j \in \{i+1, ..., n\}$, $a(i)_j$ is a best response to $a(i)_{-j}$ and $b(i)_j$ is a best response to $b(i)_{-j}$, see Chap. 2.

[11] Take, for example, player 2 in the game $\Gamma(2, 1, 1)$ of example 4.8.

an instance of cooperative, subgame-perfect, and linking-proof strategy profile for the game $\Gamma(2,1,1)$.

Example 4.9 (4.8 continued). Consider again the game $\Gamma(2,1,1)$ with $d > 0$, and linking-proof networks g^* and g^1, shown by Fig. 4.4. Let the game be repeated T-times and let t^0 and t^1 be integers such that $0 < t^0 < t^1 < T$. See Fig. 4.5 for the illustration of the following strategy profile:

- On the outcome path:
  - network g^* is established in all periods,
  - player 1 cooperates during periods $\{1,..,T-t^0\}$, and
  - players 2 and 3 cooperate during periods $\{1,..,T-t^1\}$.
- If player 1 defects early, network g^1 is established and all players defect during the remaining periods.
- If player 2 or 3 defects early, network g^* is kept and all players defect during the remaining periods.

In words, the threatened loss of links during $\{T-t^0+1,...,T\}$ constitutes an incentive to player 1 to cooperate during $\{1,...,T-t^0\}$. The incentive to players 2 and 3 to cooperate during $\{1,...,T-t^1\}$ is the possibility to free ride on player 1 during $\{T-t^1+1,...,T-t^0\}$. Players will conform to the strategy profile if one-period profit of any early defection is offset by the loss of profit in the resulting path. All threats are credible as they consist of a repetition of a static equilibrium.

More precisely, let $\Delta_0 = t^0$ and let $\Delta_1 = t^1 - t^0$. If players 2 and 3 follow the strategy profile they are rewarded in Δ_1 periods by free riding on player 1, thus earning $\Delta_1 f$ instead of $\Delta_1 d$. The reward increases with its length and there certainly exists a positive integer Δ_1 such that early deviation does not pay. Similarly, if player 1 deviates early she is punished by exclusion for at least Δ_0 periods, earning 0 instead of earning at least $\Delta_0 2d$. The punishment increases with its length and there exists a positive integer Δ_0 such that early deviation does not pay. Find such Δ_0 and Δ_1, and set $t^0 = \Delta_0$, $t^1 = t^0 + \Delta_1$ and $T > t^1$. The resulting strategy profile is subgame perfect, linking-proof, and cooperative.

It is interesting to note that "intervals" Δ_0 and Δ_1 need not be very long. For example, with the prisoner's dilemma game of Fig. 2.1 it is sufficient to set $\Delta_0 = 2$ and $\Delta_1 = 1$. That is, if the game $\Gamma(2,1,1)$ is repeated at least 4 times the profile above induces all players to establish g^* and cooperate during all but the last three periods.

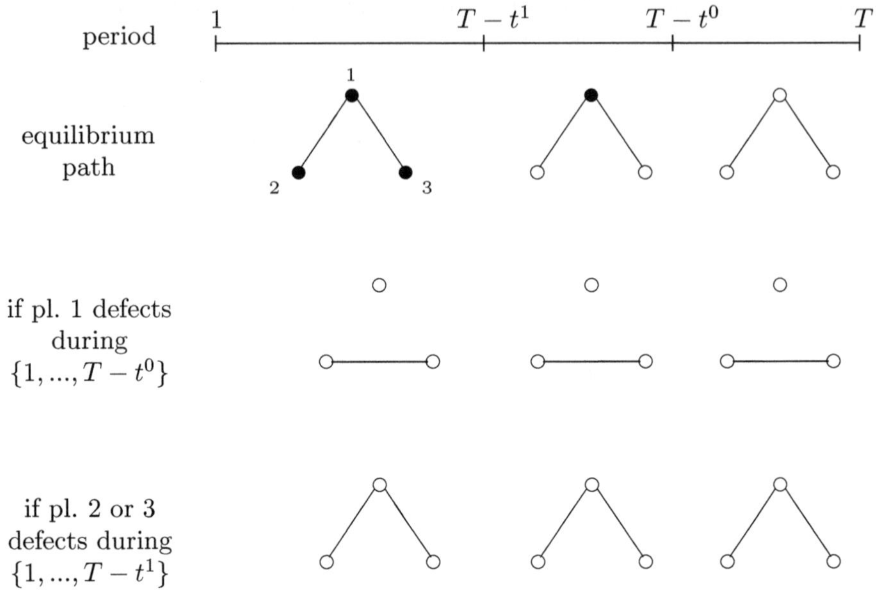

Fig. 4.5. An illustration of a subgame perfect, cooperative and linking-proof strategy profile for the repeated game of example 4.8, with $n = 3$, $\mathbf{k} = (2, 1, 1)$, and $d > 0$. All players cooperate (*dark circles*) in the initial periods and defect (*light circles*) in the final periods. Players 2 and 3 free-ride on player 1 in some penultimate periods

In the example above only one iteration of rewarding is needed to sustain complete cooperation during early periods of the repeated game. More than one iteration may be necessary for dilemma games with general linking constraints. The definition of the recursive trigger profile below suggests that the number of necessary iterations is related to the structure of the initial network structure.[12]

Let $d > 0$ and let g^* be a connected linking-proof network for $\Gamma(\mathbf{k})$. Define

$$N^0 = \{i \in N \mid l_i(g^i) < l_i(g^*) \text{ for some } g^i \in \mathcal{G}(\Gamma(\mathbf{k}))\}$$

to be the set of players who can be punished (for an early defection) by unilateral exclusion from g^*. Assume that N^0 is non-empty. Consider now the set N^1 of their neighbors,

[12] More precisely, the number of necessary iterations coincides with the maximal distance between a player who can be excluded from the network and a player who cannot be excluded.

$$N^1 = \left(\bigcup_{i \in N^0} L_i(g^*) \right) \setminus N^0 .$$

Players in N^1 can be rewarded for cooperation with the opportunity to free ride on players in N^0. Now define recursively the sets $N^2, ..., N^m$ by

$$N^\eta = \left(\bigcup_{i \in N^{\eta-1}} L_i(g^*) \right) \setminus \left(N^{\eta-1} \cup N^{\eta-2} \right)$$

and let N^m be the last nonempty set in the sequence.[13] By definition, each set N^η consists of players whose shortest distance to a player from the set N^0 is η. Given that g^* is connected, any player is at a finite distance from any other player. Each player thus belongs to some N^η. Hence, the family $\{N^\eta\}_{\eta=0}^m$ is a partition of the set of players N. For each player i define $\eta(i)$ to be such that $i \in N^{\eta(i)}$.

Definition 4.3. *Consider an n-player dilemma game $\Gamma(\mathbf{k})$ with $d > 0$. Let g^* be a connected linking-proof network and let $\{N^\eta\}_{\eta=0}^m$ be the corresponding partition of N. Assume that N^0 is non-empty. For each $i \in N^0$ let g^i be a linking-proof network such that $l_i(g^i) < l_i(g^*)$. A **recursive trigger strategy profile** for $\Gamma^T(\mathbf{k})$, based upon $(g^*, \{g^i\}_{i \in N^0}, (t^\eta)_{\eta=0}^m)$, $0 < t^0 < ... < t^m < T$, is a strategy profile denoted by $\varrho(g^*, g^*, \{g^i\}_{i \in N^0}, T, (t^\eta)_{\eta=1}^m)$ and given by*

outcome path: establish g^* in all periods, player i cooperates during $\{1, ..., T - t^{\eta(i)}\}$ and defects otherwise,
threat for N^0: if a player $i \in N^0$ defects early, i.e. during $\{1, ..., T - t^0\}$, all players defect and establish g^i during the remaining periods,
threat for N^η: if a player $i \notin N^0$ defects early, i.e. during $\{1, ..., T - t^{\eta(i)}\}$, all players defect and establish g^* during the remaining periods,
simultaneous deviations: if several players simultaneously defect early exercise the threat for the deviating player with the lowest index.

The recursive trigger profile is clearly linking-proof. It is also cooperative because all players cooperate in a connected network during $\{1, ..., T - t^m\}$. The following theorem characterizes sufficient conditions for existence of a subgame perfect recursive trigger profile.

[13] Such set exists because the number of players is finite and all sets L^η with $\eta < m$ are nonempty.

Theorem 4.1. *Consider a dilemma game $\Gamma(\mathbf{k})$ with $d > 0$. Let g^* be a connected linking-proof network. If $l_i(g^i) < l_i(g^*)$ for some player i and some linking-proof network g^i then there exists a positive number T^* such that for any integer $T > T^*$ a subgame perfect recursive trigger profile for $\Gamma^T(\mathbf{k})$ can be constructed that is (i) subgame perfect, (ii) linking-proof, and (iii) cooperative, selecting the outcome $(\mathbf{C}, g^*)$ during the periods $1, ..., T - T^*$.*

Proof. The complete proof is given in Appendix A. However, we provide here an outline of the proof.

All threats of the recursive trigger profile are credible, because they consist of a repetition of a static equilibrium. We thus need to verify that there exist positive integers $t^0, ..., t^m$ such that all threats are effective.

From the perspective of a player i, such that $0 < \eta(i) < m$, the outcome path of a recursive trigger strategy profile $\varrho(g^*, g^*, \{g^i\}_{i \in N^0}, T, (t^\eta)_{\eta=0}^m)$ can be divided into four parts as follows: (p1) during periods $\{1, ..., T - t^{\eta(i)+1}\}$ player i and all her neighbors cooperate; (p2) during periods $\{T - t^{\eta(i)+1} + 1, ..., T - t^{\eta(i)}\}$ her neighbors in $N^{\eta(i)+1}$ defect, but player i and her other neighbors continue to cooperate; (p3) during periods $\{T - t^{\eta(i)} + 1, ..., T - t^{\eta(i)-1}\}$ player i defects and free-rides on her neighbors in $N^{\eta(i)-1}$ who still cooperate; and finally (p4) during periods $\{T - t^{\eta(i)-1} + 1, ..., T\}$ player i and all her neighbors defect.

If player i defects during (p1) or (p2) then all her neighbors (and all other players) defect in all subsequent periods, and i thus looses the payoff she would otherwise have earned by free-riding on some of her neighbors during (p3). This loss increases with the length of part (p3),

$$\Delta_{\eta(i)} = (T - t^{\eta(i)-1}) - (T - t^{\eta(i)}) = t^{\eta(i)} - t^{\eta(i)-1} , \tag{4.7}$$

in a linear manner. Hence, player i will not defect during (p1) or (p2) if the length of part (p3) is so large that no unilateral deviation of player i in part (p1) or (p2) exceeds the potential loss of payoff during part (p3). We show in Appendix A that, given a finite length of part (p2), a *finite* minimal length of period (p3), $\Delta^*(i)$, exists for each i such that the threat of punishment for any her unilateral deviation is effective. No player in N^η can profit from deviating during her part (p1) if

$$\Delta_\eta \geq \max_{i \in N^\eta} \Delta^*(i) . \tag{4.8}$$

A player $j \in N^m$ faces only the parts (p1), (p3) and (p4) but not part (p2). Therefore, there exists the *finite* minimal length of period (p3), $\Delta^*(j)$, such that the threat of punishment for any her unilateral deviation is effective, independently of T and all t^η. No player in N^m can profit from deviating during her part (p1) if (4.8) is satisfied for $\eta = m$.

Given that $\Delta^*(j)$ for each $j \in N^m$ is finite, there exists a finite Δ_m that satisfies 4.8 for $\eta = m$. Given a finite Δ_m there exists a finite $\Delta^*(i)$ for each $i \in N^{m-1}$, and thus a finite Δ_{m-1} that satisfies 4.8 for $\eta = m - 1$. Using this argument recursively, we prove in Appendix A that a sequence $(\Delta_\eta)_{\eta=1}^m$ of finite integers exists such 4.8 is satisfied for each $\eta \geq 1$.

A player $i \in N^0$ faces only the parts (p1), (p2) and (p4) but not part (p3). If she deviates during (p1) or (p2) she is subsequently punished by exclusion during the remaining periods. This threat is effective if the number of periods during her part (p4), $\Delta_0 = t^0$ is sufficiently long. We prove in Appendix A that, given a finite Δ_1 a finite Δ_0 exists such that no player $i \in N^0$ can profit from deviating during her parts (p1) or (p2).

Given this $\Delta_0 = t^0$ and the finite $(\Delta_\eta)_{\eta=1}^m$ that satisfy (4.8), the sequence $(t^\eta)_{\eta=0}^m$ can now be obtained by recursive application of (4.7), such that $\varrho(g^*, g^*, \{g^i\}_{i \in N^0}, T, (t^\eta)_{\eta=0}^m)$ is subgame perfect for any $T > t^m$. □

Conditions of Theorem 4.1 are much weaker than those of Proposition 4.8. Those of Proposition 4.8 require that a network is constructed such that each player can be punished by exclusion. In contrast, conditions of Theorem 4.1 require only a connected network such that at least one player can be punished by exclusion. Proposition 4.10 below shows that for a vast majority of linking constraints *each* connected linking-proof network satisfies conditions of Theorem 4.1.

Proposition 4.10. *Consider an n-player dilemma game $\Gamma(\mathbf{k})$ with $d > 0$. If (i) $k_i \geq 2$ for each i and (ii) $k_j \leq n - 2$ for some j then connected linking-proof networks exist, and each satisfies conditions of Theorem 4.1.*

Proof. Given in Appendix A.

The condition that network g^* be connected is not necessary for the conclusions of Theorem 4.1. If g^* is disconnected then the sufficient condition for existence of a subgame perfect recursive trigger profile is that at least one player from each connected component of g^* belongs to N^0. That is, in each connected component of g^* it must be possible to punish at least one player by exclusion. To construct a cooperative, linking-proof and subgame perfect profile one may then apply the construction of a strategy profile, outlined in the proof above, to each of the connected components.

4.5 Discussion

4.5.1 Summary of the Results

In this chapter we consider finitely repeated n-player games in which each player in each period selects her interaction partners and chooses whether to cooperate or to defect. By means of individual partner choice an interaction network is established in each period. A player interacts with her immediate neighbors: she plays a prisoner's dilemma game with each of them. We call this game a network dilemma game. Players may be constrained in the number of partners they can select, in which case we refer to a game with linking constraints. If no reference to linking constraints is made we refer to general network dilemma games, with or without linking constraints.

We say that cooperation can be *sustained* if there exists a subgame perfect equilibrium such that in one or more periods: (i) no player is isolated and (ii) all players cooperate. We say that a network dilemma game Δ *permits* cooperation if cooperation can be sustained in Δ^T for some finite T.

Result 4.1. COOPERATION AND ENDOGENOUS NETWORK FORMATION *Cooperation can be sustained in many finitely repeated prisoner's dilemma games with endogenous formation of the interaction network. In contrast, there is no cooperation in finitely repeated prisoner's dilemma games played on an exogenously fixed network.*

An n-player prisoner's dilemma game played on a fixed network has a unique Nash equilibrium (see Sect. 2.2.4): each player defects. Following Lemma 2.1-2, if the game is repeated finitely many times this implies a unique subgame perfect equilibrium: all players always defect. On the other hand, if players have the possibility to modify the network the stage game may have different Nash equilibria. This is a necessary condition to sustain non-equilibrium outcomes, such as cooperation, in the repeated game. To demonstrate that cooperation can be sustained if partner choice and, consequently, the interaction network are endogenous, we have given in this chapter a number of examples of cooperative subgame perfect equilibria for the finitely repeated network dilemma games.

The possibility for endogenous partner choice is, however, not sufficient to sustain cooperation. To begin with, a network dilemma game must be repeated sufficiently many times. For example, there are no cooperative relations in any one-shot dilemma game, or in the last period of any repeated dilemma game. Next, the payoff for mutual defection should be sufficiently high: players should be willing to establish some links even if they all defect. Finally, linking constraints may be crucial. For example, a network dilemma game with unconstrained linking may not permit cooperation. Alternatively, if linking constraints are too strict cooperation may be limited only to subgroups of players.

Result 4.2. OUTSIDE OPTION
It is difficult, and in most games impossible, to sustain cooperation when the outside option is high. If $d < o$, then in any Nash equilibrium the empty network is established in each period of the repeated network dilemma game.

We have shown that the set of subgame perfect equilibria crucially depends on whether the outside option is low or high. Namely, all relations in a one-shot network dilemma game are defective. If $d < o$ then each player prefers to opt out from all defective relations. The one-shot game may thus have only one equilibrium network: no links are established. By backwards induction, the empty network is established in each period of any finitely repeated game. This is the case for all network dilemma games considered.

In the results 4.3 and 4.4 we assume that the outside option is low, $d > o$. Let $\mathbf{k} = (k_1, ..., k_n)$ be the vector of linking constraints.

Result 4.3. LINKING CONSTRAINTS
Let $d > 0$. A network dilemma game with linking constraints permits cooperation if constraints are neither extremely severe nor extremely lax. In particular, given a sufficient number of periods a linking-proof, cooperative, subgame perfect equilibrium exists if $2 \leq k_i$ for all i and $k_j < n - 1$ for some j. If $k_i = 1$ for each i (matching model) then such equilibrium exists if and only if n is odd. No such equilibrium exists if $k_i = n - 1$ for each i (unconstrained linking).

Support for this result is given by Propositions 4.9 and 4.10. To assure that punishments never rely on active participation of the punished player we require that strategy profiles are linking-proof. In absence of linking-proofness condition cooperation is possible also in games with no linking constraints (see Sec. 4.4.1).

The final result explains when cooperation can be sustained solely by threats of exclusion. For this we can use trigger strategies: any deviation by defection triggers partial or complete exclusion of defectors from the network. Trigger strategy profiles are defined by (4.5). The following is a summary of Proposition 4.8.

Result 4.4. PUNISHMENT WITH EXCLUSION
Let $d > 0$. If the number of players is relatively large compared to the number of links that players can establish then each player can be punished by exclusion using trigger strategies. Linking-proof trigger strategy profiles can sustain cooperation if $2 \leq k_i \leq \sqrt{2n-9}$ for each i.

It is important to note that the conditions in results 4.3 and 4.4 are sufficient, but most likely not necessary. We have thus characterized only a part of network dilemma games that permit cooperation. The complete characterization seems to be a challenging combinatorial task and the ideas in this chapter may find use in future search for an improved characterization. Our results nevertheless serve to advance the following main conclusions:

- *cooperation can be part of a subgame perfect equilibrium of a finitely repeated prisoner's dilemma game if it is played on an endogenous network,*
- *to achieve cooperation some competition for partners may be necessary, caused for instance by strict constraints on the number of links,*
- *cooperation can be sustained solely via exclusion of defectors if the number of all players is substantially larger than the number of links that players can support.*

4.5.2 Beyond the Network Dilemma Game

When real-world situations are modeled by simple games this typically leads to crude approximations of the richness and complexity of analyzed situations. A typical game is a stylized model of interaction emphasizing some particular real phenomenon. Obviously, when abstracting from a real situation one should seek to avoid over-simplifications, especially if there is the danger that they drive the results. Unavoidably, more realism often leads to increased complexity and may compromise analytical tractability. To address this concern we discuss the assumptions made in the course of the design and analysis of the network dilemma games in this chapter.

Through the design of the game we assume that discrimination in the action choice across different neighbors is not permitted, and that information about past actions of all players, and of the whole network structure, are common knowledge. In our analysis we assume that players are rational, that the payoffs represent their preferences, and that this is also common knowledge.

In addition, we consider in this chapter only the equilibria in pure strategies. The reason is that deviations from mixed strategies cannot always be detected and so equilibria in mixed strategies are difficult to enforce.[14] In this chapter, in fact, we suggest to restrict, rather than relax, the set of strategies in repeated network dilemma games. In particular, we consider restriction to linking-proof strategy profiles and to trigger strategy profiles. Our results suggest that in most of the games either there are no cooperative subgame perfect equilibria or there exists a linking-proof one. In this sense the restriction of linking-proofness is mild. The restriction to trigger strategies may, however, substantially reduce possibilities for cooperation.

Trigger strategy profiles were considered because they seem more intuitive than recursive trigger strategies: if a player defects too early, she is excluded from the network. However, even trigger strategies may be unrealistic if they require global transformations of the network, that is, if players need to transform the whole existing network in order to exclude one player. Future research may identify network dilemma games in which cooperation is sustained only via local transformation of the network: to exclude a player only her neighbors change their links.

In Chap. 3 we give some examples of real-world situations that may be modeled by the network dilemma game. In each example the agents must either act cooperatively in all their interactions or in none. However, there are also many situations where agents are free to simultaneously cooperate with one interaction partner, defect with another, and refuse to interact with a third agent. This seems especially true for social interaction between people. It would thus be interesting to analyze an alternative model, in which players may discriminate in the action choice across their interaction partners. A few conclusions about such an alternative model can be made using our results. Suppose that players in a network dilemma game are free to discriminate in their action choice across their neighbors, and that information about past moves, including the actions of each player in interactions with each of her neighbors, is complete. Defection of any player on any of her neigh-

[14] Smith [79] proposes a workaround by assuming a public randomization device: in every period players can condition their moves on the outcome of some publicly observed exogenous random variable.

bors is then observed by everyone and can trigger exclusion, just as in games without discrimination in actions. The only addition to strategies described in this chapter is that defection in any relation counts as deviation. This implies that, *given complete information, the set of network dilemma games that support cooperation is the same whether or not discrimination across neighbors in respect to actions is permitted.*

Relaxations of the complete information assumption are more difficult to address. In order to exclude a player that defected, her defection should be known among players that are supposed to adjust their links. The information condition thus limits the possibilities for exclusion by permitting only the transformations of the network that involve players which are aware of the defection. For example, if a player knows only the actions of her neighbors, exclusion can take place only via local transformation of the network.

Special cases of information conditions may be particularly interesting for future research. One such condition is that the actions of a player are known only by her neighbors. Another condition, in which a network structure may play an important role, is that actions of a player are known also by neighbors of her neighbors, that is, there is a limited flow of information through the network, e.g. through gossip.

Information about the network structure could also be limited, which certainly is the case for social networks. Previous research has shown that people are surprisingly successful in guessing network characteristics even when they know only their local neighborhood (see [89, 90, 91, 92]). In relation to the network dilemma games, a local knowledge of a network may be sufficient to induce exclusion of defectors. For example, if a player is aware of the links of her neighbor she may be able to make conjectures about her cooperation: if the neighbor suddenly looses links it may be that she defected with another player, is hence unreliable, and should be excluded. Again, further research should look into the effects for cooperation of various relaxations of the complete information assumption.

Finally, our assumptions about common knowledge of rationality and preferences are in tradition with standard non-cooperative game theory. However, previous research has shown that, especially in social dilemma games, theoretical results may change if any of these assumptions is relaxed even slightly. See e.g. [14] for relaxation of the common knowledge assumption, [93, 94] for models with boundedly rational players, and [95] for a model with other-regarding players.

In particular, it is unlikely that these traditional assumptions are satisfied in experiments. For example, in experiments the game is imposed upon the subjects using monetary payoffs, which may not correctly represent player's preferences. Therefore, in order to discuss our experimental results, we analyze in Chap. 5 the solutions of the repeated network dilemma game played by other-regarding players. Furthermore, in Chap. 6 we discuss an alternative model of behavior in repeated network dilemma game, assuming boundedly rational players.

4.6 Conclusion

It is well known that a finitely repeated prisoner's dilemma game has a unique subgame perfect equilibrium: players defect in each period. The same holds for n-player prisoner's dilemma games played in fixed groups or on a fixed network. In this chapter, however, we prove that a different result holds if the network is endogenously generated by the players. In particular, we show that *cooperation can be sustained in a subgame perfect equilibrium of a finitely repeated prisoner's dilemma game played on an endogenous network.*

We construct such equilibria under the traditional assumptions of common knowledge of rationality, complete information and selfishness. The main rationale behind our result is that players can be discouraged from defection if they fear to be punished by exclusion from the interaction network. In particular, threats with exclusion can be credible if players are constrained in the number of links. In such cases players can organize themselves into different equilibrium networks. For example, the equilibrium network may be such that one player is isolated, thus foregoing payoffs from playing the game.

We identify sufficient conditions for the existence of cooperative subgame perfect equilibria for finitely repeated prisoner's dilemma games on endogenous networks. These depend on the game length, the outside option value and the linking constraints. Introducing endogenous network formation itself is not sufficient for cooperation, but assuming, in addition, very weak constraints on the number of links may be sufficient. This assumption is satisfied in many real situations, for example in social networks where people are constrained in the number of their acquaintances.

5

Exclusion and Cooperation in Experiments

Laboratory experiments are a useful way of gathering observations about human decision-making in controlled environments. A careful experimental design can give valuable indications about human decision-making in economic environments outside the laboratory. Experimental data is in general replicable and is less prone to problems of missing observations and measurement errors, compared to empirical data collected in the field. The experimental method is especially useful to explore dynamics of behavior in situations for which empirical observations are otherwise difficult to collect, for example, the dynamics of behavior in social and economic networks.

In this chapter we use the experimental method to study the co-evolution of cooperation and partner choice in various network dilemma games with mutual link formation. Our main conclusions are that subjects exclude defectors and establish cooperative networks both in presence and in absence of linking constraints. A benchmark experiment where a prisoner's dilemma is played on a fixed network confirms that cooperation dramatically increases when subjects are free to choose their interaction partners.

In the following section we outline our approach and the position of our experiment in view of the related experimental literature. The details of our experimental design and our research questions are given in Sect. 5.2. Experimental procedures are given in Sect. 5.3. In Sect. 5.4 we list the theoretical equilibrium predictions using the analysis in the previous chapter, as well as our alternative hypotheses that are based on the results of other related experiments. We list, motivate and discuss our results in Sect. 5.5. In Sect. 5.6 we apply two models of social preferences to characterize equilibria of the games in our experiment when played by other-regarding players. We measure the other-regarding preferences of the participants in our experiment and investigate whether their cooperation can be related to their preferences. We summarize in Sect. 5.7.

5.1 Introduction

The experimental evidence about behavior in social dilemma games is rather mixed. On the one hand, experiments with prisoner's dilemma games typically report substantial levels of cooperation, with averages significantly above the theoretical prediction of no cooperation. Similarly, the average contributions in experimental public goods game often significantly exceed the theoretically predicted level of uniform free-riding. These observations encourage the view that the free-rider problem can be overcome without the exogenous support of institutional enforcements.

On the other hand, the evidence from experiments with *repeated* social dilemma games does not support this view. In repeated multi-player prisoner's dilemma games most subjects eventually defect, and contributions in repeated public goods games steadily decrease toward the theoretically predicted level of free-riding. One exception is the repeated two-player prisoner's dilemma game in which cooperation is often sustained by subjects who using tit-for-tat types of strategies [19].

As a consequence of the diverse evidence the social dilemmas remain among the most investigated topics in experimental economics. A recent major advance came with the experimental study by Fehr and Gächter [96] who demonstrate that contributions in the experimental public good game can be sustained at high levels if subjects have the possibility to punish other players. In contrast with the theoretical prediction they observe a widespread willingness among subjects to incur costs to punish free-riders. Crucially, they also report that this willingness does not disappear over periods, which is necessary for the observed steady increase in contributions. This observation matches the general experimental evidence that many subjects behave reciprocally [97, 98, 99, 100, 101, 102], and that frequency and strength of reciprocal behavior do not decline over time [99, 103, 104].

Punishment is a mechanism that gives reciprocal types of people an opportunity to retaliate, and thus reduces free-riding. We show in this chapter that partner choice is another such mechanism. This is important because direct punishment may not be possible in all situations. For example, inflicting physical harm or destroying another person's property is generally against the law. In contrast, avoiding unwelcome interaction may quite often be possible and is socially acceptable.

Exclusion can be seen as a sort of punishment albeit quite different from the direct punishment studied for instance in [96]. Direct punishment is assumed to have an immediate effect and a sure loss for both the punisher and the punished player. It has no direct effect for future interaction. In contrast, the loss associated with exclusion is uncertain because it implies a loss of future interaction, the value of which depends on future behavior. This indicates an important relation between partner choice, the resulting network structure, and the spread of norms of behavior. For example, if cooperators consistently exclude defectors the society may segregate into isolated subgroups of cooperators and defectors. The transition of an individual from a defective into a cooperative group may then be difficult and individuals may instead adopt the cooperative standards of their local neighborhood. As a consequence, free-riding may never disappear from the population.

The effect of exclusion of defectors depends on the loss associated with separation from the interaction network. In an experiment with public goods games with punishment, Egas and Riedl [105] demonstrate that contributions increase only when punishing is cheaper than being punished. In the setting of our network dilemma games, if outside options are the same for all players then defectors always lose more from exclusion than cooperators. In this sense, cooperators are encouraged to exclude defectors.

We consider mutual link formation because this seems to be the natural case. We do not impose any linking costs and conjecture that some reciprocal subjects are willing to exclude defectors even in the absence of linking constraints and at a cost of the reduced number of interactions. We consider two experimental treatments. In one the subjects engage in a network dilemma game without any linking constraints. In the other treatment we impose strict linking constraints. In addition, we investigate the effect of endogenous choice of links by comparing the experimental treatments with a control treatment in which we fix the interaction network exogenously.

Our results can be summarized as follows. Cooperation increases in all experimental treatments, compared to the control treatment in which cooperation declines. The impact of linking constraints on the magnitude of cooperation appears limited. However, in contrast with the theory presented in the previous chapter cooperation seems to *increase* with rounds when linking is unconstrained and decrease when linking is strictly constrained. Moreover, groups that face linking constraints tend to segregate into subnetworks of cooperators and defectors. Such segregation is much less evident in groups that have no linking constraints.

Our experiment is one of the first to investigate the simultaneous dynamics of networks and behavior. In fact, the literature on experiments with network formation began to emerge only recently and only a couple of papers have been published (see [106, 107]). A review of this literature is given in [108]. In line with the theoretical literature, most of the existing experiments in economics focus only on network formation without any parallel dynamics of behavior on the network (see e.g. the review by Jackson [41]). Network experiments have a longer tradition in sociology and social psychology, but focus primarily on the relation between network position and bargaining power (see e.g. [109, 110, 111]). The most similar to ours is the experiment of Corbae and Duffy [46] who study the evolution of coordination in endogenous networks. They study repeated coordination games on networks that can be periodically modified by the players and observe that networks usually converge to a collection of isolated pairs. The remaining directly related experiments focus on multiple prisoner's dilemma games with outside option. Orbell and Dawes [25] study one-shot prisoner's dilemma games with high outside option and report increased cooperation between players who chose to play the game. Hauk [112] confirms this observation in the repeated setting but reports decreased cooperation in the game with intermediate outside option. Hauk and Nagel [24] report that cooperation in games with high outside option can be further increased if linking is unilateral (see also [113]). None of these three papers, however, investigates the dynamics of interaction structures.

Experimental evidence shows that the possibility of migration increases cooperation. Commitment in a buyer-seller environment sustains cooperation in presence of incomplete contracts when termination of a relationship is used as a disciplining device [114]. Contributions to a public good can be increased through endogenous reorganization of groups [115]. Interestingly, cooperation can increase also with an exogenously imposed exclusion or reorganization of groups [116, 117]. On the other hand, efficiency does not increase if cooperators cannot prohibit defectors from infiltrating their groups [118].

Finally, the structure of interaction itself may influence the dynamics of cooperation in prisoner's dilemma games on exogenously imposed networks. Kirchkamp and Nagel [22] observe lower cooperation on circles than in fully connected groups. Cassar [21] observes that cooperation increases with segregation, but remains difficult to achieve. Both experimental studies report decreasing cooperation levels with averages below 50 percent.

In contrast to the above literature we explicitly study the dynamics of the network structures and its relation to the dynamics of cooperation. Furthermore, we identify the effects for this relation of environmental variables such as the possibility to choose partners and the linking constraints. We demonstrate that these effects can be large and may determine whether groups segregate or integrate. We also conclude that, regardless of the other environmental effects, the endogenous choice of partners may be sufficient to sustain cooperation.

5.2 Experimental Design

Our design consists of two experimental and one control treatment. In each treatment we study groups of 6 subjects playing 60 periods of an experimental game. The composition of groups and the identities of subjects in each group do not change across periods.

In each of the experimental treatments we consider a network dilemma game with mutual link formation and the outside option $o = 0$. In the control treatment we consider a 6-player prisoner's dilemma game played on an exogenously fixed network. The experimental games in all three treatments are based on the same prisoner's dilemma game, with the following payoff matrix:

	cooperate	defect
cooperate	50, 50	10, 70
defect	70, 10	30, 30

(5.1)

The two experimental treatments differ only in the constraints imposed on the maximal number of links that a subject can propose:

I - UNCONSTRAINED: linking is not constrained; each subject can propose up to 5 links.

II - CONSTRAINED: each subject can propose up to 2 links.

In the control treatment III - CONTROL the complete network is exogenously imposed, and subjects cannot choose or modify their links. In each period each subject plays a prisoner's dilemma game with all five other subjects in her group.

If in treatment UNCONSTRAINED we remove the possibility to endogenously choose the links and fix the network to the complete network, then the resulting game coincides with the experimental game in the CONTROL treatment. Hence, the control treatment is directly comparable to the experimental treatment I with no linking constraints.

The motivation for our choice of treatments is as follows. First, we consider mutual link formation, which means that a proposed link is established only if it is reciprocated. This gives each subject a permanent possibility to avoid interaction with any other subject. In other words, mutual linking enables each subject to exclude any other subject regardless of that subject's linking and action choices. Second, the theoretical considerations in Chap. 4 suggest that the set of linking-proof subgame-perfect equilibria of a network dilemma game depends crucially on whether or not linking is constrained. We therefore consider two experimental treatments, one with and another without the linking constraints. Finally, in order to detect the impact of partner choice for cooperation we complement the two experimental treatments with the control in which we take away the freedom to propose or modify one's own links.

5.2.1 Research Questions

Our first research question concerns the effect endogenous linking has on cooperation. We conjecture that cooperation is easier to sustain when partner choice is endogenous. In the control treatment a subject can retaliate defection by another subject only by defecting herself. This, however, may trigger retaliation by even more players and lead to uniform defection. In contrast, in the experimental treatment a subject can retaliate also by removing her links with defectors. The subject can signal cooperative intentions and at the same time punish defectors. This may sustain cooperation and even induce defectors to cooperate. By comparing behavior in each of the experimental treatments with behavior in the control treatment we therefore investigate *whether endogenizing the network structure promotes cooperation.*

The standard equilibrium analysis, presented in Chap. 4, predicts no exclusion in treatments with unconstrained linking. In absence of opportunity costs, when the outside option value is zero, establishing a link increases one's payoff regardless of the actions of the linked individuals. Removing a link is, therefore, always costly and individuals will exclude defectors only when they are exogenously constrained in the number of their links. Free-riding may, consequently, decrease only under linking constraints. In contrast with this conclusion, the evidence from related experiments, described in Sect. 5.4, suggests that people are willing to punish defectors and manage to sustain cooperation even when punishment is costly. Hence, our second and third research questions are *do subjects exclude defectors when exclusion is costly* and *do linking constraints increase cooperation.*

Finally, by analyzing behavior in each group we can study the relation between the social structure and cooperation. For instance, if linking behavior of cooperators substantially differs from that of defectors then cooperators and defectors should have different neighborhood structures. Further, if cooperators are more likely to have cooperative neighbors than defectors then cooperation may turn out to be profitable. Our fourth research question is *how neighborhoods and earnings differ between cooperators and defectors.*

5.3 Experimental Procedures

All sessions of the experiment were conducted in the CREED laboratory at the University of Amsterdam. In total 96 subjects participated. Each experimental session lasted between 90 and 120 minutes. Average earnings were 20.9 euros. Subjects were students of various disciplines but mainly economics and business administration undergraduates. They were recruited through notices on bulletin boards in the faculties and through email announcements. Each subject participated in only one session and none had participated in a similar experiment before. To ensure anonymity at least twelve subjects were recruited for every session, and were randomly divided into at least two independent groups. All sessions ran as planned and without any problems. We had 6 groups in treatment I and 5 groups in each of the treatments II and III. Each group is one independent observation.

Each session consisted of two parts. The first part consisted of a test of the subject's other-regarding preferences. In the second part the subjects played 60 periods of the experimental game. To minimize the influence between the two parts the subjects received no information about the decisions of other subjects in the first part until the end of the experiment. Furthermore, the decision of subjects in the first part did not affect their earnings in the second part. We therefore consider the results from the two parts as independent.

Before the first part of the experiment subjects were explained the rules of conduct and the structure of the whole experiment, followed by the detailed instructions for this part. They were told that another part of the experiment will follow, that their decisions in the first part will not be revealed until the end of the whole experiment and that these decisions have no effect for the payoffs in the second part. Only after all subjects finished the first part did they received instructions for the second part of the experiment. After reading the instructions each subject had to answer a number of questions, designed to test their understanding of the network game, the payoff calculations and the computer screen. After all subjects correctly answered all questions there was a single practice period allowing subjects to ask questions about making decisions via the computer screen, at the end of which no information about the choices of other subjects was revealed. The second part of the experiment started only after all subjects confirmed they had no more questions.

The experiment was computerized using software developed in CREED by Jos Theelen as a general tool for network experiments. All instructions were read aloud and shown on the computer screen. Each participant also received a printed summary of instructions immediately before each part of the experiment began. The test questions and the practice period for the second part were computerized as well.

A detailed description of the first part of the experiment is given in Sect. 5.6.4, where we discuss the relation between the observations in the two parts. In Sects. 5.4–5.5 we focus on the second part of the experiment. We next describe the procedures for this part.

No communication other than through decision-making was permitted. To increase anonymity and avoid suggestive framing each subject referred to herself as "Me" and to the other five members of her group with the letters "A", "B", "C", "D", and "E". Each letter referred to the same other subject throughout all periods, and this was common knowledge. The network dilemma game was neutrally framed: subjects were forming "connections" and choosing between the actions "green" and "blue". Care was taken to minimize differences in the instructions for different treatments. In all treatments the subjects were informed about the size of their group, procedures to propose and establish links, the payoff function, and the number of periods. Appendix B contains the precise experimental instructions that were given to the subjects.

An example of the computer screen seen by the subjects in the second part of the experiment is shown in Fig. 5.1. The screen is divided into a left and a right side. Subjects make decisions on the right side of the screen and receive information about the history of play on its left side. Linking choices are made by clicking on the buttons "A", "B", "C", "D", and "E" in the window "decision: connections". In the treatment CONSTRAINED the subjects are not able to propose more

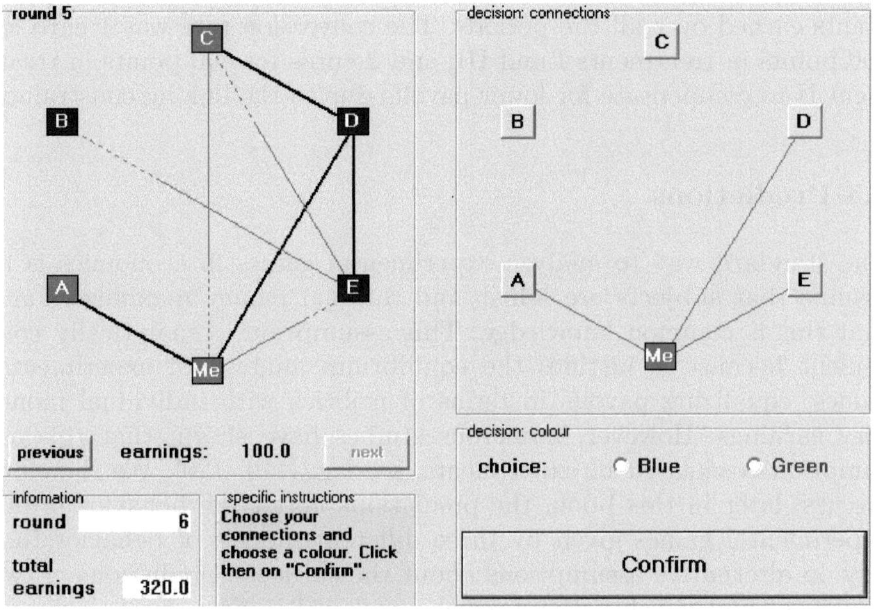

Fig. 5.1. Computer screen for the experimental network dilemma game

than two links simultaneously. In the treatment CONTROL the subjects cannot propose any links. Instead, all links are considered as proposed in each period.

An action is chosen by selecting either the color "blue" (cooperate) or "green" (defect) in the window "decision: color". When the subject is satisfied with her decision she clicks on the "confirm" button. After all subjects made their decisions the information about the proposed and established links and the chosen colors is revealed in the window "round #". Established links are shown as thick black lines. Links that were proposed but not reciprocated are shown as thin gray lines, broken on the side of the subject that did not propose the link. All proposed and established links are visible and each letter is colored with the color chosen by the corresponding subject. Each subject is always able to access the complete history of play. The buttons "previous" and "next" are used to browse through the information that the subject received in previous periods.

In accordance with the tradition of experimental economics the payoffs earned in both experimental parts translate into monetary earnings, paid individually to each subject in cash at the end of the experimental session. The payoff of a subject was determined by the sum of the points earned over all the periods. The conversion rate was 1 euro for 500 points in treatments I and III, and 2 euros for 500 points in treatment II to compensate for lower payoffs due to the linking constraints.

5.4 Predictions

The standard way to analyze experimental games in economics is to assume that subjects are selfish and rational money maximizers and that this is common knowledge. This assumption is analytically convenient because it justifies the equilibrium analysis of experimental games, equalizing payoffs in terms of utilities with individual monetary earnings. However, numerous studies have shown that this assumption is violated in experiments, see e.g. [119, 120]. We therefore discuss, later in this book, the predictions about the behavior in our experimental games given by three different models of behavior that rely on alternative assumptions about the subjects: predictions of two models of non-selfish preferences are analyzed in Sect. 5.6 below, and in Chap. 6 we look at the dynamics of behavior in network dilemma games played by boundedly rational players. In this section, however, we discuss predictions based on standard equilibrium analysis, and alternative predictions that are based on previous experimental evidence and intuition.

5.4.1 Measures of Cooperation and Network Structure

We shall distinguish between the level of cooperation and the level of cooperative play. The level of cooperation is the number of cooperative actions in a group in one period and does not take the network structure into account. The level of cooperative play, on the other hand, measures cooperation in games that were actually played and thus weights each subject's action with the number of her links.

Take any profile of moves (a, p) of the 6 subjects in one group. Let δ_i be 1 if subject i cooperates and 0 otherwise and let $l(p)$ be the number of all established links. The *level of cooperation* is the normalized number of subjects who choose to cooperate, regardless of their linking decisions:

$$\textit{level of cooperation} = \frac{1}{6}\sum_{i \in N} \delta_i \ ,$$

If all subjects in the group cooperate the level of cooperation is 1; if all subjects defect this level is 0. The *level of cooperative play* measures the rate of cooperation in the games that were actually played: it is calculated as the number of cooperative actions over all links, normalized by double the number of established links:

$$\textit{level of coop. play} = \frac{1}{2l(p)} \sum_{g(p)_{ij}=1} (\delta_i + \delta_j) \ ,$$

This measure is not affected by actions of subjects that are isolated from other subjects in the established network. Further, the weight of a subject's action increases with the number of games she plays, that is, with the number of her neighbors. The level of cooperative play is 1 if all non-isolated subjects cooperate and 0 if they all defect. The two measures coincide if the network is complete (as e.g. in the control treatment).

The *density* of the network, $g(p) = l(p)/15$, gives the number of established links normalized by 15 (the number of all possible links between 6 agents is 15). The density is thus 1 when the network is complete and 0 when the network is empty. The number of links proposed by subject i, $\sum_{j \in N} p_{ij}$, is her *outdegree*. The indegree of subject i is the number of other subjects that propose a link to i, $\sum_{j \in N} p_{ji}$.

5.4.2 Standard Equilibrium Predictions

Assume that the utility of each subject coincides with her individual monetary earnings, that each subject is rational and that this is common knowledge. Let $N = \{1,2,3,4,5,6\}$, let the vector of linking constraints in treatment II be given by $\mathbf{k}^\circ = (2,2,2,2,2,2)$, and let the prisoner's dilemma payoff function v° be defined by (5.1). The experimental games then coincide with the following network dilemma games:

Treatments I and II: $\Gamma^I = \langle N, J, \pi \rangle$ and $\Gamma^{II} = \langle N, J(\mathbf{k}^\circ), \pi \rangle$, where

$$\pi_i(a,p) = \sum_{j \in L_i(p)} v^\circ(a_i, a_j),$$

Treatment III: $\Gamma^{III} = \langle N, A, \pi^c \rangle$, where $\pi_i^c(a) = \sum_{j \in N \setminus \{i\}} v^\circ(a_i, a_j)$.

Using the theory developed in Chap. 4 we get the following characterization of equilibria of our finitely repeated experimental games.

Treatment I - UNCONSTRAINED: In each period all players establish all links and defect, in the unique linking-proof subgame perfect equilibrium. However, cooperation can be sustained in a non-linking-proof subgame perfect equilibrium, for example with strategies "Cooperate until period 59 and defect in period 60. Remove the links with all other players that defected in the previous period and propose all other links. If you deviated in the previous period then remove all your own links". Cooperation can also be sustained in a non-subgame perfect Nash equilibrium, for example with strategies "Propose all links in period 1. Cooperate until period 59 and defect in period 60. Remove the links with all other players that defected in the previous period and propose all other links".

Treatment II - CONSTRAINED: Cooperation can be sustained in a linking-proof subgame perfect equilibrium, for example with strategies "Cooperate until period 59 and defect in period 60. Propose your two links to those two other players that cooperated in the previous period and are closest to square 'Me' on your screen. If you yourself defected then defect again and propose your two links to those two other players that defected in the previous period and are closest to square 'Me' on your screen. If less than two players cooperated in the previous period then defect and propose your two links to players 'A' and 'E' on your screen". Other linking-proof subgame perfect equilibria exist. In some, for instance, all players defect in each period.

Treatment III - CONTROL: In the unique Nash equilibrium all players defect in each period.

See Results 4.1 and 4.3 in Chap. 4. The results about the uniqueness of linking-proof subgame-perfect equilibria for treatment I can be derived as follows. The complete network is the unique linking proof network of the stage game. By definition of a linking-proof subgame perfect equilibrium in Chap. 4, this uniqueness implies that the complete network is established in each period, along and off the equilibrium path, of any such equilibrium. All players defect in the last period of the finitely repeated game independently of the history. Hence, both the actions and the linking choices of all players in the last period are independent of the history. All players then defect also in the penultimate period. We can now repeat the argument and use backwards induction to show that, along the outcome path of the linking-proof subgame-perfect equilibrium of the finitely repeated game, all players establish the complete network and defect in each period.

Based on the equilibrium analysis above we can formulate the following predictions. In treatment I all links will be established and all players will defect in each period. In the control treatment III all subjects will defect in each period. The analysis is ambiguous about behavior in treatment II.

5.4.3 Alternative Predictions

The experimental game in the control treatment III can be interpreted as a public goods game with a binary choice [121, 35]. Previous experimental evidence shows that contributions in repeated public goods games initially substantially exceed those predicted by the standard equilibrium analysis, but that in subsequent periods the contributions gradually decrease toward the predicted uniform free-riding [122, 123, 124]. A similar downward trend from positive initial levels of cooperation can, therefore, be expected in our control treatment.

This downward trend can be reversed into an upward trend if each period of the public goods game is supplemented by a "punishment" phase in which subjects can reduce the payoffs of other subjects at some cost to themselves [96]. Many subjects are willing to incur costs to punish defectors but contributions increase across periods only if the effect of punishment, i.e. the reduction in payoff of the punished subject, exceeds the cost of punishing [105]. While the games in our experimental treatments are different from public goods games with punishment, there is some analogy. First, individual exclusion may be used to punish defectors in prisoner's dilemma games. This may have a similar positive effect for cooperation that individual punishment has for contributions in public goods games. Second, exclusion of defectors by cooperators is costly both for the excluding and for the excluded subjects, because they both forego a sure positive payoff from mutual interaction. However, the cooperator's cost of excluding another player is never higher than the cost of the excluded player. Third, the standard equilibrium predictions coincide in our experimental games with low outside option and those on a fixed network, just as they coincide in public goods games with and without punishment. Motivated by this analogy we predict, for our unconstrained experimental treatment I that some cooperators will punish defectors with exclusion and that cooperation rates will increase across periods.

Ostracism can increase cooperation when imposed exogenously [116]. Similarly, players cooperate more when they compete for the opportunity to play the prisoner's dilemma game [125]. In each case cooperation is achieved through exclusion of the least cooperative players. In the

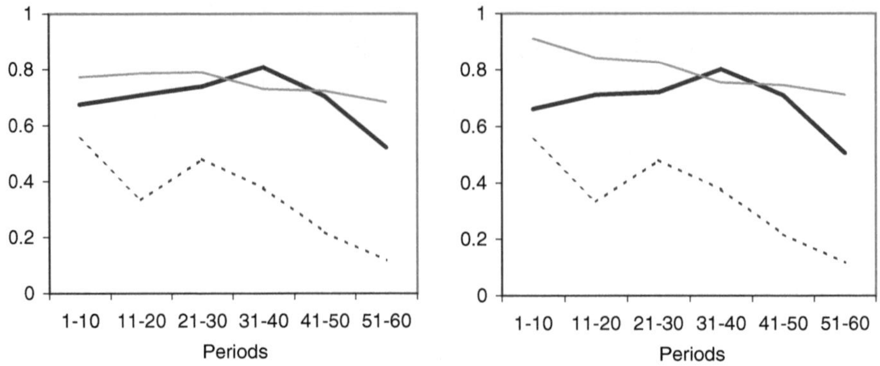

Fig. 5.2. Dynamics of cooperation rates in treatments I-unconstrained (*thick line*), II-constrained (*thin line*) and III-control (*dashed line*), each aggregated in clusters of 10 periods. The left panel displays cooperative choice rates and the right panel displays cooperative play rates

setting of our network dilemma games, the competition for partners can be increased through the introduction of linking constraints. We therefore predict that, among our three treatments, the exclusion of defectors and the cooperation rates will be highest in the constrained experimental treatment II.

To summarize, we predict a downward trend in cooperation in treatment III and an upward trend in cooperation in treatments I and II. In all treatments the players earn more from a cooperative than from a defective neighbor. Hence, subjects in treatments with endogenous link formation may be more likely to propose links to those other subjects that they believe will cooperate. We therefore predict that cooperators have a higher indegree and establish more links than defectors. Our final prediction is that segregation between cooperators and defectors will be highest in treatment II.

5.5 Experimental Results

Before we discuss the details of our results we outline our main observations. Fig. 5.2 shows condensed illustrations of the evolutions of cooperation and of cooperative play for the different treatments. As expected, in none of the treatments the theoretically predicted one-shot equilibrium play is observed right from the beginning. However, there are remarkable differences in the dynamics of cooperation, in particular between the control treatment with an exogenously fixed network and the two experimental treatments with endogenous linking. Cooperation exhibits a downward trend in the control treatment and, with the exception of the final periods, is high and stable or even exhibits an upward trend in each of the experimental treatments. Linking constraints appear to increase cooperation in the treatments with endogenous linking only in the initial and the final periods.

An impression about our subject's linking behavior and cooperation can be gained from the illustrations of average cooperation and average network structures, shown in Figs. 5.3–5.5. In these figures each circle represents one subject in the group. The ratio of cooperative actions of that subject is represented by the colored fraction of the circle. The style of the line connecting any two circles represents the incidence of an established link between the corresponding two subjects. The interpretation of the different line styles is given in Table 5.1.

The networks in treatment I - UNCONSTRAINED differ from the theoretically predicted complete networks. However, in line with the predictions of the theory in Chap. 4, the networks in treatment II - CONSTRAINED are linking-proof as subjects seem to establish as many links as possible. Moreover, in this treatment the cooperative subjects appear to successfully separate from the defective subjects, which leads to a segregation of groups into two subnetworks – one almost completely cooperative and another where cooperation is less frequent. There seems to be a relation in both experimental treatments between individual cooperation and the number of established links: individuals that cooperate relatively more have relatively more neighbors.

We now proceed with our detailed results using various measures of cooperation, linking behavior and the network structure, and of their dynamics.

Result 5.1. COOPERATION IN EXOGENOUS AND ENDOGENOUS NETWORKS
First-period cooperation rates are around or above 0.5 in all three treatments. During periods 1-55 the cooperation rates converge to uniform defection only in the control treatment III with the exogenous network. There is no such tendency during periods 1-55 in the experimental treatments with endogenous networks: in treatments I and II the cooperation rates stabilize around 0.7. In periods 31-59 the cooperation rates in the experimental treatments with endogenous link formation significantly exceed those in the control treatment. There is a substantial end-game effect in both experimental treatments: cooperation rates drop to between 0.1 and 0.4 in period 60.

Table 5.1. The interpretation of line styles in Figs. 5.3–5.5, illustrating the incidences of established links

Line style	Number of periods (#) in which the link was established
thick line	$51 \leq \#$
thin line	$41 \leq \# \leq 50$
dashed line	$21 \leq \# \leq 40$
dotted line	$11 \leq \# \leq 20$
no line	$\# \leq 10$

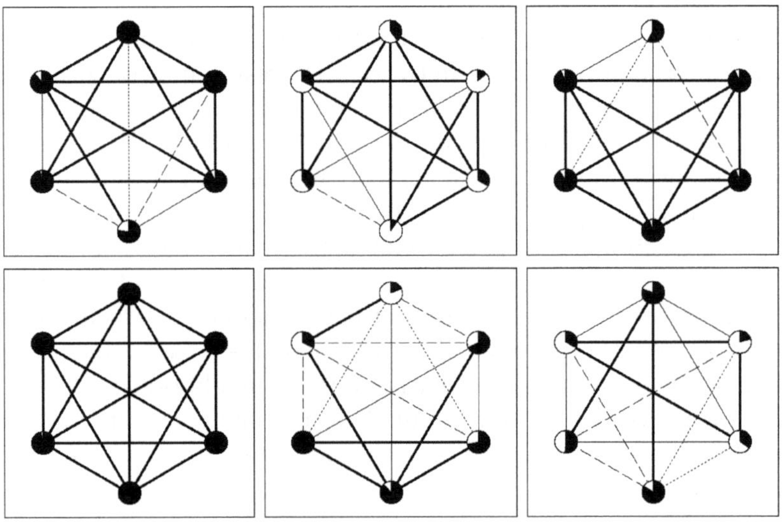

Fig. 5.3. Average individual cooperation and average network structures in treatment I-unconstrained

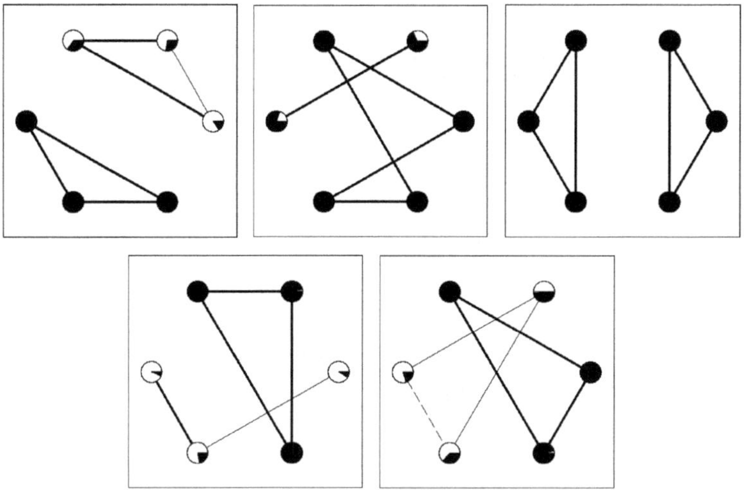

Fig. 5.4. Average individual cooperation and average network structures in treatment II-constrained

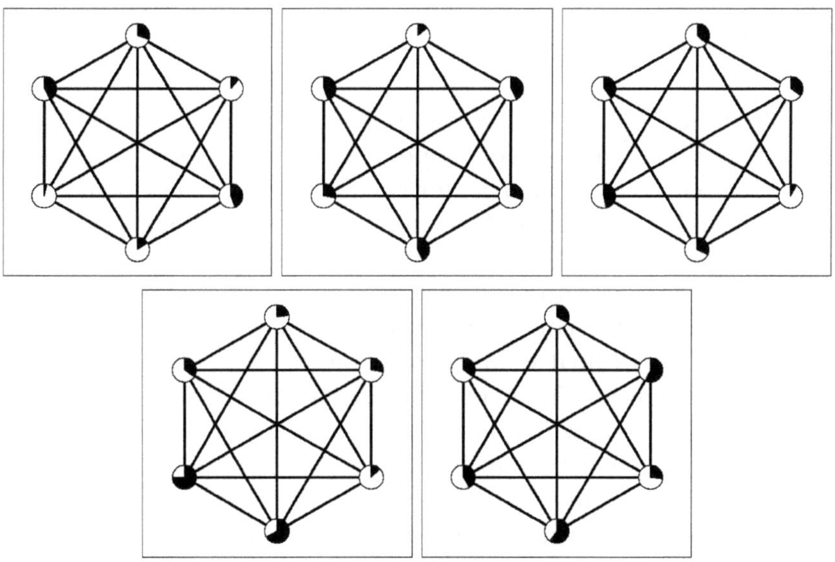

Fig. 5.5. Average individual cooperation in treatment III-control

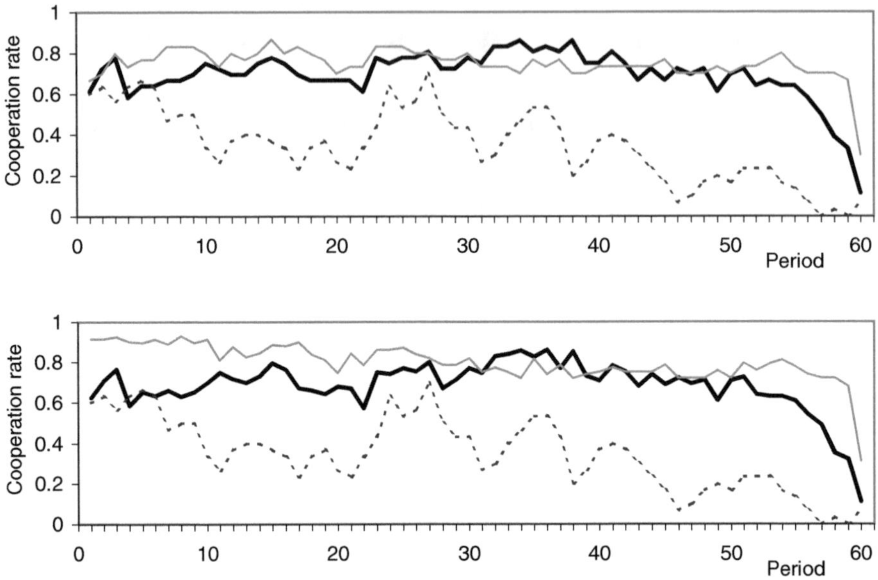

Fig. 5.6. Dynamics of cooperation rates in treatments I-unconstrained (*thick line*), II-constrained (*thin line*) and III-control (*dashed line*). The upper panel displays cooperative choice rates and the lower panel displays cooperative play rates

The first support for this result comes from Fig. 5.6 which depicts, for each treatment, the rates of cooperative choices (*upper panel*) and of cooperative plays (*lower panel*), for each of the sixty periods. Observe that in the first period, there is a substantial amount of cooperation in all treatments (see also Table 5.2). What these figures also show is that the dynamics of cooperation differ significantly between the experimental treatments and the control treatment. In particular, cooperation in treatment III - CONTROL exhibits a declining trend on both measures. Interestingly, there are frequent attempts to increase cooperation (see the 'hills' in Fig. 5.6) which are, however, not successful. In contrast, both experimental treatments exhibit very high and stable cooperation rates across periods, on both measures. The cooperation rates in treatment I increase until they reach a maximum in periods 31 to 40, temporarily exceeding cooperation rates in treatment II, which remain high until almost the final period. Thereafter, due to the end-game effect, cooperation rates decrease again. The average rates of cooperative play in treatment II-CONSTRAINED exceed those in treatment I-UNCONSTRAINED in the initial and the final periods. However, these differences are not statistically significant, as we report next.

Table 5.2. Cooperation rates and their standard deviations (*in brackets*) across all periods and across periods 1-55. Treatment averages across groups

	Cooperative choices:		
Treatment	first period	periods 1-60	periods 1-55
I - unconstrained	0.611	0.693	0.721
		(0.188)	(0.140)
II - constrained	0.667	0.748	0.760
		(0.127)	(0.106)
III - control	0.600	0.346	0.375
		(0.269)	(0.262)

	Cooperative plays:		
Treatment	first period	periods 1-60	periods 1-55
I - unconstrained	0.626	0.686	0.715
		(0.198)	(0.150)
II - constrained	0.917	0.798	0.813
		(0.136)	(0.109)
III - control	0.600	0.346	0.375
		(0.269)	(0.262)

To get further support for the above result we ran non-parametric Mann-Whitney tests for the first period, periods 1-10, 11-20, 21-30, 31-40, 41-50, 51-60, and the last period, with group averages in each treatment as units of observation. These statistics corroborate the observations drawn from the figures. The first and last period actions do not differ between the three treatments.[1] In later periods, however, cooperation is higher in both experimental treatments I and II compared to the control treatment III. In particular, in periods 31-40, 41-50, and 51-60 the differences are significant at least at the 6 percent significance level (all tests one-sided and corrected for multiple comparisons). Cooperation rates do not differ between the two experimental treatments. This holds for both measures of cooperation. The same results are found when we look at the aggregated cooperation levels across all periods or (to control for the end-period effect) across periods 1-55 (see also Table 5.2). A Mann-Whitney test shows that aggregated cooperative choices and cooperative plays in treatments I and II significantly exceed those in treatment V (at the 5 percent level, one-sided and corrected tests).

[1] One group in treatment II did not establish any links in the first period, reducing the number of observations of cooperative play in the first round of this treatment to four, thus prohibiting meaningful statistical testing.

An examination of Fig. 5.6 also suggests that the dynamics of behavior in experimental treatments exhibit much more stability than in the control treatment. To find statistical support for this impression we calculate the standard deviation of cooperative choices and plays across the periods for each group. Table 5.2 shows in parentheses the averages of these standard deviations per treatment for periods 1-60 and for periods 1-55. For cooperative choices a Mann-Whitney test shows that standard deviations in treatment III significantly exceed those in treatments I and II (at the 2.5 percent level for all pair-wise comparisons, one-sided and corrected tests) for both ranges of periods and both measures of cooperation. Together with Result 5.1 this shows that the possibility to form the network endogenously, i.e. to choose with whom to play, serves as a powerful mean for promoting *and* stabilizing cooperation at a high level.

Result 5.2. STABILITY OF COOPERATIVE BEHAVIOR
The action choices in treatments with endogenous network formation exhibit significantly more stability compared to the control treatment.

We now turn our attention to the evolution of network structures. We begin with the analysis of the dynamics of the network density. Recall that the network in the control treatment III is fixed to the complete network with its density equal to 1. The (linking-proof) equilibrium prediction for treatment I with unconstrained linking is also the complete network with density of 1. The density in treatment II with constrained linking can reach at most 0.4. Such density is achieved in a linking-proof network in which each subject establishes two links.

Result 5.3. NETWORK DENSITIES
In experimental treatment I the network density is rather high in all periods (around 0.8) although substantially below its theoretically predicted value of 1. In treatment II the network density steadily increases towards its theoretically predicted maximum of 0.4.

Support for this result comes from Fig. 5.7, showing the dynamics of network densities over time. The average network density (across all periods) in the treatment I with unconstrained linking is around its average of 0.82 in all periods, slightly increasing only in the last periods. For reference, the density of 0.8 means that 12 links out of the 15 possible are established. In contrast, the average network density in treatment II with constrained linking increases monotonously until the final periods. As can be seen from illustrations of the established networks in Fig. 5.4 the subjects tend to established stable linking-proof

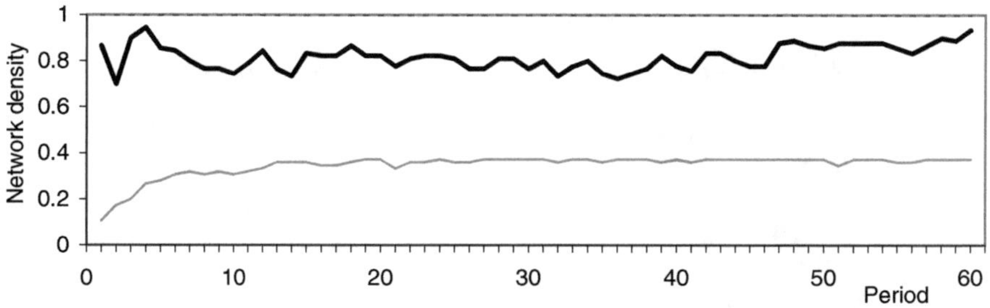

Fig. 5.7. Dynamics of average network densities in treatments I-unconstrained (*thick line*) and II-constrained (*thin line*)

networks in four of the five groups in this treatment. The proportion of linking-proof networks in treatment II with constrained linking is 0.67, while this proportion in treatment I with unconstrained linking is only 0.29.

Having observed high cooperation rates in both experimental treatments it is natural to ask for the driving force behind this result. The most natural candidate is (fear of) exclusion. To study the linking and partner choosing behavior of subjects we proceed as follows. For each period $t \geq 2$ we look at links proposed by subjects that cooperate in period t (C_t). We count how frequently they propose a link to other subjects that cooperated in the previous period (C_{t-1}) and to those that defected in the previous period (D_{t-1}). This indicates, for example, how often they exclude previous-period defectors or previous-period cooperators. In the same way we calculate how frequently the links to past cooperators and defectors are proposed by current defectors (D_t). The resulting frequencies, aggregated for each experimental treatment across all subjects and all periods, are depicted in Fig. 5.8.

The figure shows that in both treatments cooperators are relatively unlikely to link to previous period defectors. In treatment I the current cooperators propose links to past defectors only in 46 percent of all possible cases. Hence, in 54 percent of all cases cooperators are willing to incur costs to exclude previous-period defectors.

Exclusion of defectors is intensified in treatment II, where cooperators propose a link to previous period defectors only in 7 percent of all possible cases. It is intuitively clear that exclusion of past defectors is more frequent in the treatment with linking constraints, as cooperative subjects prefer to propose their few links to other cooperators. In both experimental treatments cooperators propose on average 30 percent more links to previous-period cooperators than to previous-period defectors ($p < 0.03$, one-sided paired Wilcoxon signed rank tests, with

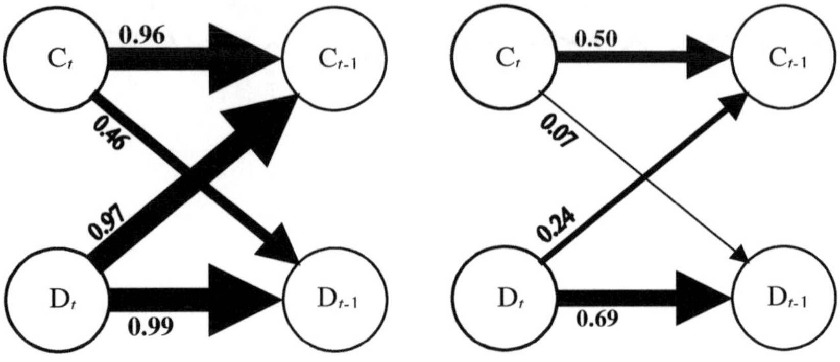

Fig. 5.8. Linking behavior in treatments I-unconstrained (*left panel*) and II-constrained (*right panel*). The thickness of an arrow is proportional to the ratio of the number of links proposed by current-period cooperators (or, depending on the arrow, defectors) to previous-period cooperators (or defectors), relative to the number of all possible such links, averaged across all subjects and all periods

groups as observation units). It is interesting, however, that we observe a difference in the defectors' linking behavior only in treatment II with constrained linking, where defectors on average propose 30 percent more links to previous-period defectors ($p < 0.03$, one-sided paired Wilcoxon signed rank test, with groups as observation units). This indicates that cooperators try to isolate defectors and, in the case of constrained linking, that defectors accept their isolation and link only to other defectors.

Result 5.4. PROPOSED LINKS
Cooperators are much more likely to propose links to previous-period cooperators than to defectors in both experimental treatments. Cooperators do not hesitate to exclude previous-period defectors even when they are free to propose and establish any number of links, that is, when exclusion is costly. When linking is not constrained (treatment I) defectors try to link to everybody. When linking is constrained (treatment II) defectors are much more likely to propose links to previous-period defectors than to cooperators.

This tells us that intending cooperators are more likely to link to past cooperators than to past defectors. Excluding *past* defectors does not ensure, however, that *present* defectors are excluded. There is the possibility that past cooperators change to actual defectors and vice versa. To examine whether cooperators were able to exclude defectors we need to look at actions in established networks.

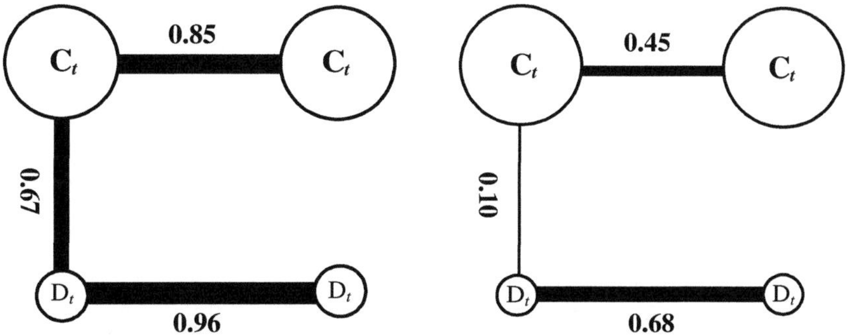

Fig. 5.9. Established links in treatments I-unconstrained (*left panel*) and II-constrained (*right panel*). The thickness of a line is proportional to the ratio of the number of links proposed by cooperators (or, depending on the line, defectors) to other cooperators (or defectors), relative to the number of all possible such links, averaged across all subjects and all periods

Result 5.5. Established links
In both experimental treatments there is a tendency toward segregation between cooperators and defectors: cooperators are more likely to have cooperative neighbors and defectors are more likely to have defective neighbors. The segregation is more pronounced when linking is constrained. In particular, in treatment II the defectors are often completely separated from cooperators.

Support for this result comes from Fig. 5.9. To obtain this figure we calculate, from all established networks per treatment, how frequently two cooperators or two defectors established a link, and how frequently was the link established between a cooperator and defector. The diameters of the circles around C_t (cooperator in period t) and D_t (defector in t) are proportional to the average rates of cooperators and defectors in the corresponding treatments.

For treatment I with unconstrained linking the figure shows that cooperators establish links with almost all other cooperators (in 85 percent of all possible cases) and defectors establish links with almost all other defectors (96 percent). The lower ratios of established links between cooperators and between defectors in treatment II are, obviously, a consequence of the linking constraints. In both treatments a link between a cooperator and defector is least likely to be established (67 and 10 percent, respectively). This shows that cooperators are not only trying to exclude defectors (see Result 5.4) but are also, at least to a certain level, successful in doing so. Additional evidence of the

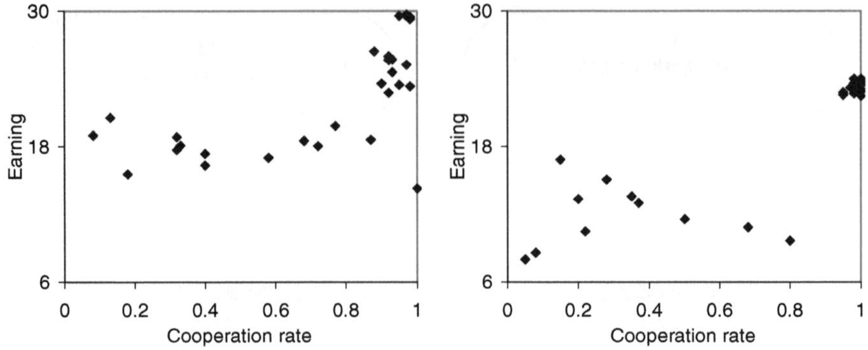

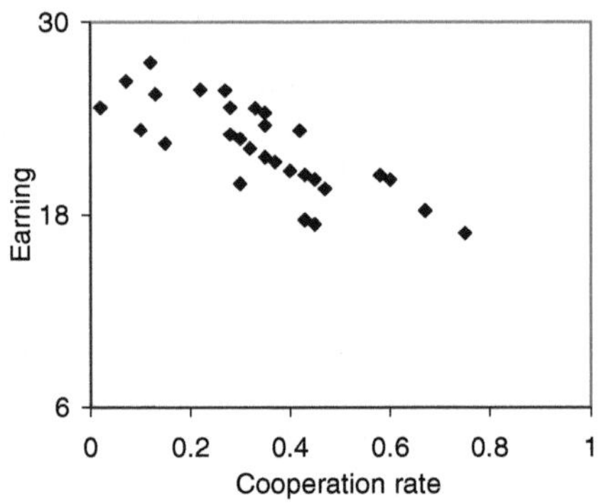

Correlation coefficients

Treatment	r	obs.	p
I	0.59	36	0.0408
II	0.87	30	0.0000
III	−0.79	30	0.0000

Fig. 5.10. Individual cooperation and earnings in treatments I-unconstrained (*upper left panel*), II-constrained (*upper right panel*) and III-control (*lower left panel*). Each point represents one subject's average cooperation and her total earnings. The Spearman rank order correlation coefficient is calculated for each treatment (*lower right panel*)

strict segregation between cooperative and defective subnetworks in treatment II can be found in the illustrations in Fig. 5.4.

The segregation between cooperators and defectors reduces the possibility for free-riding. A subject who frequently defects may lose her links with cooperative neighbors and earn less than a subject who persistently cooperates. On the other hand, no segregation is possible in treatment III with a fixed network, which makes defection more profitable than cooperation. To investigate the differences in earnings between cooperators and defectors we plot in Fig. 5.10, for each treatment, the average cooperation of each subject against her total earnings. For treatment III the plot reveals a clear negative relation between cooperation and earnings. The fact that defectors in this treatment consistently earn more than cooperators is supported by the highly negative

Spearman rank order coefficient ($r = -0.79, p < 0.001$) of correlation between cooperation and earnings. In contrast, the plots for experimental treatments I and II reveal no such negative relation. If anything, the positive and significant ($p < 0.05$ for each experimental treatment) Spearman rank order correlation coefficients ($0.59 \leq r \leq 0.87$) suggest that highly cooperative subjects, on average, earn more than highly defective subjects.

Result 5.6. COOPERATION AND EARNINGS
In treatment III there is a substantial negative correlation between cooperation and earnings. In contrast, this correlation is positive in each of the experimental treatments I and II.

5.6 Social Preferences

In the theoretical equilibrium analysis of network dilemma games in Sect. 5.4 we assume that the game payoffs correspond to player's utilities.[2] In experimental games, however, the payoffs correspond to monetary earnings and these may not correctly represent the utilities of subjects. More specific, some experimental subjects may be "other-regarding" rather than "selfish" and may care about the monetary earnings of other subjects as well as about their own. Alternatively, subjects may reward other subjects for good intentions and punish them for selfish or aggressive behavior. In terms of their preferences, experimental subjects may thus perceive the game quite differently from the game described by monetary payoffs [126].

Many recent models of human decision-making assume that a substantial number of people are motivated by concerns for fairness and reciprocity. Some of these models assume that fair-minded people dislike outcomes with unfair distribution of monetary earnings [127, 95]. Another class of models assign a major role to fair intentions [128, 129, 130, 131]. Both types of models are commonly called models of social preferences. In this section we apply two models of social preferences to analyze whether assuming fair-minded players would better explain the behavior in our experimental network dilemma games.

We consider the outcome-based model of inequity aversion by Fehr and Schmidt [95] and the model of altruistic preferences by Levine [132] that may be viewed as a simple intentions-based model. Both models

---

[2] In game theory it is always assumed that the ordering of payoffs in games correctly represent the preferences of hypothetical players: a player always prefers a higher own payoff over a lower one.

were shown to be successful in predicting behavior in public goods games with or without punishment, which are related to our network dilemma games. Their authors consider the models as static and do not intend to explain the dynamics of behavior in repeated games. Rather, they aim at predicting the convergence of behavior in experimental games that were repeated many times, such that players have a fair knowledge about preferences of the other players. We, therefore, study only one-shot games with fair-minded players and compare their equilibria to the last period behavior in the experimental games.[3]

For each of the two models we proceed as follows. We consider games obtained when our experimental games are played either by altruistic or by inequity averse hypothetical players under common knowledge of preferences. We study the equilibria of such games and identify the necessary conditions for cooperation. For each underlying experimental game we show that if altruism and inequity aversion are distributed as described by the original author's empirical application of their models then no cooperation is possible in equilibrium. We then use the results from the first part of our experiments to elicit the distribution of altruism and inequity aversion among our own experimental subjects. We conclude, by comparing these results to our experimental data, that we cannot explain the observed cooperation in the last period of our experimental games using either the model of altruism or the model of inequity aversion.

5.6.1 Notation

Consider a game in which outcomes yield monetary earnings to the players. Each of the two models of social preferences that we study below assumes that each player's preferences over outcomes can be captured by her utility function which assigns utility, valued in real numbers, to each possible profile of monetary earnings. Each also assumes that players are rational and maximize their utility.

We therefore distinguish between the monetary earnings and the utilities that players gain as the result of an outcome of a game. In particular, we define the relevant game theoretic concepts in terms of utility functions. For each profile of moves $(a, p) \in J$ let $\pi_i(a, p) \in \mathbb{R}$

[3] Given the strong end-game effect in our experimental games we also considered to compare the predictions of the two models with behavior in some earlier period, rather than with behavior in the final period. We decided to remain consistent with the approach of the authors who test their models using the data from the final period of the repeated public goods game.

be the monetary earning of player i. We refer to $\pi_i(a,p)$ as her *direct value*. For each i let

$$u_i : \mathbb{R}^n \to \mathbb{R}$$

be her utility function. For each profile of direct values $\pi = (\pi_1, ..., \pi_n)$ the *utility* of player i is given by $u_i(\pi)$. For convenience we denote the utility of player i given a profile of moves (a,p) by

$$u_i(a,p) = u_i(\pi(a,p)) \ ,$$

where $\pi(a,p) = (\pi_1(a,p), ..., \pi_n(a,p))$.

Move $(a_i^*, p_i^*) \in J_i$ is a *best response* of player i to the profile of moves $(a,p) \in J$ if

$$u_i(a_i^*, p_i^*, a_{-i}, p_{-i}) \geq u_i(a_i', p_i', a_{-i}, p_{-i})$$

for all $(a_i', p_i') \in J_i$. Profile $(a^*, p^*) \in J$ is a *Nash equilibrium* if (a_i^*, p_i^*) is a best response to (a^*, p^*) for each player i. Profile $(a^*, p^*) \in J$ is a *linking-proof equilibrium* if it is a Nash equilibrium and if for each pair of distinct players i and j either (a_i^*, p_i^*) is a best response to $(a^*, p^* \oplus ji)$, or (a_j^*, p_j^*) is best response to $(a^*, p^* \oplus ij)$.[4]

We restrict our analysis to games Γ^I (treatment I - UNCONSTRAINED), Γ^{II} (treatment II - CONSTRAINED), and Γ^{III} (treatment III - CONTROL), defined in Sect. 5.4. In all three games the prisoner's dilemma payoffs are given by the payoff function v°, defined in (5.1).

5.6.2 Altruistic Preferences

This model was proposed by Levine [132]. Let each player i be associated with her *coefficient of altruism* γ_i, satisfying

$$-1 < \gamma_i < 1 \ . \tag{5.2}$$

Player's coefficient of altruism describes her concern for other players. It affects how the direct values of all players are translated into her utility. Levine refers to the player with $\gamma_i < 0$ as *spiteful*, to the player with $\gamma_i = 0$ as *selfish*, and to the player with $\gamma_i > 0$ as *altruistic*. For each vector of direct values $\pi = (\pi_1, ..., \pi_n)$ the *utility* of player i is given by

$$u_i = \pi_i + \sum_{j \neq i} \frac{\gamma_i + \lambda \gamma_j}{1 + \lambda} \pi_j \ , \tag{5.3}$$

---

[4] See Chap. 4 for an extensive discussion of the concept of linking-proofness. In short, a linking-proof equilibrium is a Nash equilibrium in which there is no pair of separated players that would both like to establish the mutual link.

where the parameter λ, constrained by

$$0 \leq \lambda \leq 1 , \tag{5.4}$$

reflects the fact that players have a higher regard for altruistic opponents than spiteful ones.

When $\lambda = 0$ the model is one of pure altruism of the type discussed in [133]. Levine, however, estimates that $\lambda = 0.45$ is more consistent with data from ultimatum game experiments of [104]. We call coefficients and corresponding utility functions *feasible* when conditions (5.2) and (5.4) are satisfied. Levine shows that the averages of contributions observed in the final period of the repeated public goods game experiments in [134] are roughly consistent with the following distribution of coefficients of altruism:[5]

$$\gamma_i = \begin{cases} -0.9 \; ; \; pr = 0.20 \\ -0.22 \; ; \; pr = 0.52 \\ 0.133 \; ; \; pr = 0.04 \\ 0.243 \; ; \; pr = 0.17 \\ 0.453 \; ; \; pr = 0.07 \end{cases} . \tag{5.5}$$

Consider now a network dilemma game $\Gamma = \langle N, J, \pi \rangle$, where $\pi_i(a,p)$ is the direct value earned by player i given the profile of moves $(a,p) \in J$. Let

$$\gamma = (\gamma_1, ..., \gamma_n)$$

be the vector of the player's coefficients of altruism, each satisfying condition (5.2). Let the utility of player i, given the outcome $(a,p) \in J$, be

$$u_i^A(a,p) = \pi_i(a,p) + \sum_{j \neq i} \frac{\gamma_i + \lambda \gamma_j}{1 + \lambda} \pi_j(a,p) .$$

We refer to the game $\Gamma_A(\gamma) = \langle N, J, u^A \rangle$ as the *network game with altruism.*

Proposition 5.1 characterizes the minimal altruism that would induce a subject to cooperate in a Nash equilibrium of a network game with altruism based on one of our experimental games Γ^I, Γ^{II}, and Γ^{III}. The corresponding network games with altruism are denoted by $\Gamma_A^h(\gamma)$, $\Gamma_A^l(\gamma)$, and $\Gamma_A^c(\gamma)$.

For a non-isolated player i in network g let

$$\widehat{\gamma}_{-i}(g) = \frac{1}{l_i(g)} \sum_{j \in L_i(g)} \gamma_j$$

be the average coefficient of altruism of her neighbors in g.

[5] We use pr as an abbreviation for probability.

Proposition 5.1. *Assume* $0 \leq \lambda \leq 1$ *and* $-1 < \gamma_i < 1$ *for each* i. *In any Nash equilibrium* (a^*, p^*) *of* $\Gamma_A^I(\gamma)$, $\Gamma_A^{II}(\gamma)$ *or* $\Gamma_A^{III}(\gamma)$ *with* $g^* = g(p^*)$ *a non-isolated player* i *cooperates only if*

$$\gamma_i \geq \frac{1+\lambda}{2} - \lambda\widehat{\gamma}_{-i}(g^*) \tag{5.6}$$

and defects only if the inequality sign is rotated.

The proof is given in Appendix A. Note that the assertion of the above Proposition holds for any $n \geq 2$. If the coefficients of altruism were constrained, as suggested by Levine's estimates (5.5), by $-1 < \gamma_i < 0.5$, then it can be shown that the inequality (5.6) cannot be satisfied. [6]

5.6.3 Inequity Aversion

The following model was proposed by Fehr and Schmidt [95]. Let each player i be associated with two *coefficients of inequality aversion* α_i and β_i, satisfying

$$0 \leq \beta_i < 1 \text{ and } \beta_i \leq \alpha_i \ . \tag{5.7}$$

Coefficient α_i describes player i's aversion to disadvantageous inequality, such as envy. Coefficient β_i describes her aversion to advantageous inequality, such as guilt. For each vector of direct values $\pi = (\pi_1, ..., \pi_n)$ the *utility* of player i is given by

$$u_i(\pi) = \pi_i - \frac{\alpha_i}{n-1}\sum_j \max\{\pi_j - \pi_i, 0\} - \frac{\beta_i}{n-1}\sum_j \max\{\pi_i - \pi_j, 0\} \ . \tag{5.8}$$

This model assumes that people may suffer from inequality: if inequality among players increases then the utility of a player with positive coefficients of inequality aversion decreases. The second of conditions (5.7) captures the idea that people suffer less from inequality if it is to their advantage. We call coefficients and corresponding utility functions *feasible* when they satisfy conditions (5.7). Fehr and Schmidt use experimental data to estimate the distribution of coefficients of inequity aversion by

$$\alpha_i = \begin{cases} 0 \ ; \ pr = 0.3 \\ 0.5 \ ; \ pr = 0.3 \\ 1 \ ; \ pr = 0.3 \\ 4 \ ; \ pr = 0.1 \end{cases} , \quad \beta_i = \begin{cases} 0 \ ; \ pr = 0.3 \\ 0.25 \ ; \ pr = 0.3 \\ 0.6 \ ; \ pr = 0.4 \end{cases} . \tag{5.9}$$

[6] Levine's estimates imply $\widehat{\gamma}_{-i}(g^*) < 0.5$, translating inequality (5.6) into $\gamma_i \geq 0.5$.

Fehr and Schmidt show that for this distribution the model of inequity aversion is roughly consistent with the experimental data from the public goods game with punishment, reported in [96].

Consider a network dilemma game $\Gamma = \langle N, J, \pi \rangle$, where $\pi_i(a,p)$ is the direct value earned by player i given the profile of moves $(a,p) \in J$. Let

$$\alpha = (\alpha_1, ..., \alpha_n) \text{ and } \beta = (\beta_1, ..., \beta_n)$$

be the vectors of the player's coefficients of inequity aversion, satisfying conditions (5.7). Given the outcome $(a,p) \in J$ let the utility of player i be

$$\begin{aligned} u_i^F(a,p) = \pi_i(a,p) &- \frac{\alpha_i}{n-1} \sum_j \max\{\pi_j(a,p) - \pi_i(a,p), 0\} \\ &- \frac{\beta_i}{n-1} \sum_j \max\{\pi_i(a,p) - \pi_j(a,p), 0\} \ . \end{aligned}$$

We refer to the game $\Gamma_F(\alpha, \beta) = \langle N, J, u^F \rangle$ as the *network game with inequity aversion.*

Proposition 5.2 characterizes the minimal β-coefficient that would induce a subject to cooperate in a Nash equilibrium of a network game with inequity aversion based on one of our experimental games Γ^I, Γ^{II}, and Γ^{III}. The corresponding network games with inequity aversion are denoted by $\Gamma_F^h(\gamma)$, $\Gamma_F^l(\gamma)$, and $\Gamma_F^c(\gamma)$.

Proposition 5.2. *Assume* $0 \leq \beta_i < 1$ *and* $\beta_i \leq \alpha_i$ *for each* i. *In any Nash equilibrium of* $\Gamma_F^I(\alpha, \beta)$, $\Gamma_F^{II}(\alpha, \beta)$ *or* $\Gamma_F^{III}(\alpha, \beta)$ *a non-isolated player* i *cooperates only if* $\beta_i \geq \frac{n-1}{n+1}$.

The proof is given in Appendix A. Note that the assertion of the above Proposition holds for any $n \geq 2$. For $n = 6$, it implies that $\beta_i > 0.7$ holds for each cooperative subject i. This is not possible if inequity aversion is distributed by 5.9.

5.6.4 Other-Regarding Preferences of our Experimental Subjects

In the first part of each of our experimental session the subjects were anonymously paired and participated in the following task, called the circle test. Each subject had to choose, simultaneously and without any interaction with other subjects, an angle from the interval $[-\pi, \pi]$. By choosing an angle τ_i subject i earned $500 * \cos \tau_i$ points and the other subject in her pair earned $500 * \sin \tau_i$ points. After both parts of the experiment have been completed the points earned in the circle test were converted into monetary earnings, at the exchange rate of 1 euro for 500 points. Subject i maximized her own earning with $\tau_i = 0$. Selecting $\tau_i = \frac{\pi}{4} = 45°$ maximized the joint earning of the players in the pair. In addition to the points that the subject allocated to herself she also earned the points that were allocated to her by the subject she was paired with.

The angle chosen by a subject is an indicator of her attitude toward an anonymous other subject. A subject who cares only about her own earning should choose $\tau_i = 0$. If, however, she is willing to forego a fraction of her own earning in order to increase the earning of another subject then the subject should select an angle from the interval $(0, \frac{\pi}{2}]$. A subject may also gain utility by reducing the earning of another subject, in which case she selects an angle from the interval $[-\frac{\pi}{2}, 0)$.

Higher angles within the interval $[-\frac{\pi}{2}, \frac{\pi}{2}]$ are, therefore, associated with higher concern for the other player. The circle test differs from usual dictator games, in which a player simply divides a certain amount of money between herself and another player, in that "monetary" efficiency continuously changes with the angle and is maximized for $\tau = \frac{\pi}{4}$.[7] With the help of this circle test it is possible to infer the distribution of individual coefficients of altruism, or of coefficients of inequality aversion, from the observed distribution of chosen angles. In particular, we show below that if subjects make decisions according to the models of altruism or inequity aversion, most of the chosen angles reveal their coefficients of altruism or inequality aversion.

Fig. 5.11 shows the computer screen for this task. Subjects make their decision by clicking on the circle in the left part of the screen. A vector is drawn with each click, pointing to where the subject has clicked. Two buttons below the circle can be used for fine adjustments of the vector. The angle τ between the horizontal line next to the "Me" sign and the vector, clockwise, determines the point earnings of the decision-making subject ($\cos\tau$) and of the other subject in her pair ($\sin\tau$). Any choice on the right side of the circle yields positive own earning for the subject. A choice on the upper side of the circle yields positive earning for the other, paired subject. Each time the vector is moved the point earnings are re-calculated and shown on the bottom right part of the screen. When the subject is satisfied with her decision she clicks on the "confirm" button.

[7] See, for instance, [135, 136] for description and discussion of experimental dictator games.

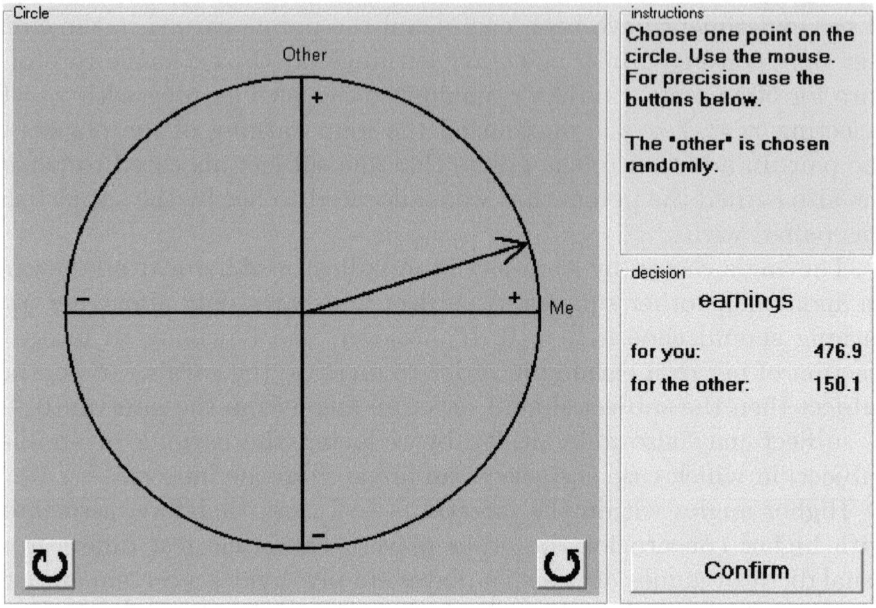

Fig. 5.11. Computer screen during the circle-test

Crucially, during the circle-test the subjects did not have any information about the subsequent experimental treatment. They were told only that another part of the experiment would follow their current task. The subjects were not informed about the angles chosen by the other subjects until the end of the whole experiment. They also knew that their choices in the circle-test would not be made public during the experiment. We therefore analyze the data from the circle-test as independent from the subsequent experimental treatments.

A thorough discussions of the circle-test are given in [125] and [137]. The circle-test is an adaptation of another test, called the ring-test, which is used by social psychologists to measure attitude toward an unknown, generalized other (see e.g. [138, 139]). In the ring-test the subject is asked to make 24 choices between different pairs of earnings for herself and a random other. From these choices an angle on a circle can be computed, which represents the *value orientation* of the subject. See [140] and [141] for a detailed description of the ring-test. For applications of the ring-test in experimental economics see also [142]. The circle-test significantly simplifies the ring-test, because the angle

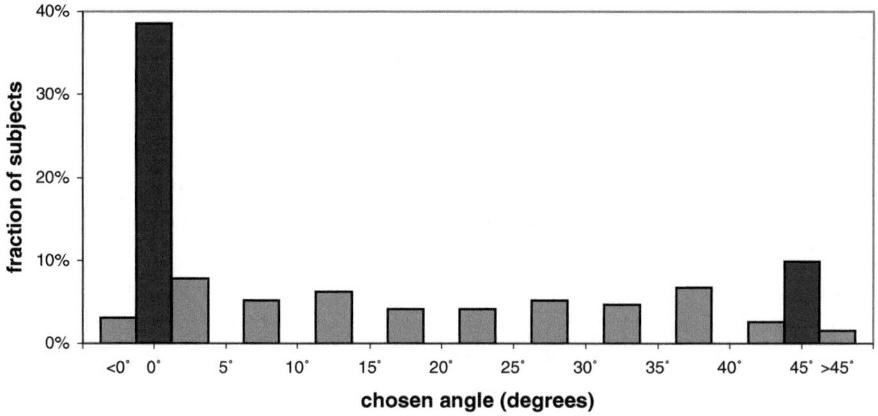

Fig. 5.12. The distribution of angles chosen in the circle-test by our subjects. The left-most column gives the proportion of subjects that chose a negative angle. The dark column "0°" gives the proportion of subjects that chose the angle 0 (0°). The dark column "45°" gives the proportion of subjects that chose the angle $\frac{\pi}{4}$ (45°). The right-most column gives the proportion of subjects that chose an angle strictly above $\frac{\pi}{4}$. Each remaining column gives the proportion of subjects that chose an angle between the two corresponding values

is chosen by the subject directly.[8] A common observation in the literature mentioned above is that both tests have predictive power: the value orientation, represented by the angle, correlates highly and positively with contributions in the initial periods of a repeated public good game.

The distribution of angles chosen in the circle-test, across all subjects in our experiment is shown in Fig. 5.12. There are two modes of choices: the angle $0 = 0°$, chosen by 38.5 percent of subjects, and the angle $\frac{\pi}{4} = 45°$, chosen by 9.9 percent of subjects. Most of the remaining choices are rather uniformly distributed across $(0, \frac{\pi}{4})$. Only 4.7 percent of all subjects chose an angle outside the interval $[0, \frac{\pi}{4}]$.

In the following subsections we describe how coefficients of altruism or inequality aversion can be computed from an angle chosen in the circle-test. We are able to calculate these coefficients for most of our subjects[9], and thus estimate the distributions of altruism and inequity aversion among them. In Sect. 5.6.5 we discuss whether we can

[8] However, consistency of the value orientation cannot be estimated by the circle-test. This is possible only by repeating the decision, as in the ring-test.

[9] Some subjects chose an angle that, according to the corresponding model, should never be chosen. Consequently, we cannot cannot make any inferences about their

explain the observed last-period cooperation levels in our treatments with other-regarding preferences of our subjects.

Coefficients of Altruism

Assume that the preferences of subject i comply with utility function (5.3) for some (unknown) coefficients λ and γ_i. Assume that i believes that the coefficient of altruism of the other player in her pair is $\overline{\gamma}_i$. We need to estimate the value of $\overline{\gamma}_i$ in order to infer the value of i's coefficient of altruism γ_i from her choice of angle in the circle test.

Our assumption about how subjects form beliefs about preferences of others is based on a prominent hypothesis in psychology, concerning expectations and behavior in social dilemma type of games. The *false consensus hypothesis* conjectures that, in absence of any information, people tend to expect that unknown others are similar to them, with regard to value orientations. The hypothesis was formulated by Kuhlman and Wimberley [143] and an extensive literature has since documented the propensity of people to expect other people to be like them [144, 145]. [10]

Hence, we assume that

$$\overline{\gamma}_i = \gamma_i \ , \tag{5.10}$$

for each subject i, that is, subjects expect that the coefficient of altruism of the unknown other player coincides with their own.[11] The preferences of subject i over the angles in the circle-test are then described by the utility function

$$u_i(\tau_i) = \cos\tau_i + \gamma_i \sin\tau_i \ . \tag{5.11}$$

Proposition A.1, given in Appendix A, describes the relation between chosen angles and coefficients of altruism. To summarize, if the utility is maximized by an angle τ_i^* then

$$\gamma_i = \frac{\sin\tau_i^*}{\cos\tau_i^*} = \tan\tau_i^* \ . \tag{5.12}$$

Angles should be in the interval $(-\frac{\pi}{4}, \frac{\pi}{4})$, and there is a one-to-one relation with coefficients of altruism.

---

coefficients. Only a small proportion of subjects chose such an angle, however: 1.6% for the model of altruism and 4.7% for the model of inequity aversion.

[10] We could have considered many alternative assumptions about subject's beliefs, such as that all subjects know the average coefficient of altruism in the subject population and respond to that when faced with an unknown other subject. However, because of the empirical support for the false consensus hypothesis, this is the only case that we consider.

[11] With this assumption the model (5.3) is equivalent to one of pure altruism, of the type discussed in [133], characterized also by $\lambda = 0$.

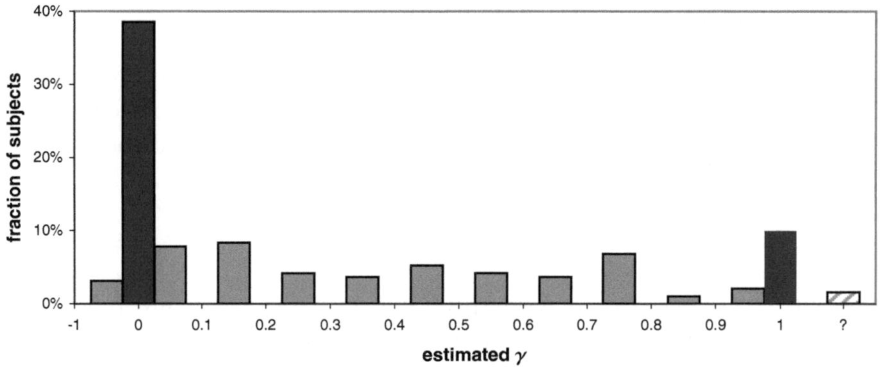

Fig. 5.13. The distribution of estimated coefficients of altruism among our experimental subjects. Dark colored columns "0" and "1" represent the proportions of subjects with $\gamma_i = 0$ or $\gamma_i = 1$. The first column represents the proportion of subjects with $\gamma_i \in (-1, 0)$. The patterned column "?" represents the proportion of subject whose choice can be explained only by $\gamma_i > 1$. The remaining columns represent the proportions of subjects whose estimated coefficient is between the corresponding two values

Using (5.12) we estimate the coefficients of altruism of our subjects. The distribution of estimated coefficients is shown in Fig. 5.13. It is evident that our distribution differs substantially and significantly ($p < 0.01$, two-sided Kolmogorov-Smirnov one-sample test) from the one given by (5.5). Around 10 percent of our subjects chose the angle $\frac{\pi}{4} = 45°$ which is consistent with $\gamma_i = 1$, and is at the boundary of the feasibility set. The average estimated coefficient among our subjects is $\widehat{\gamma} = 0.29$, which is substantially higher than -0.21, the average of distribution (5.5).

We doubt that the difference between our distribution and Levine's distribution is due to differences between subject pools. Rather, this suggests that the model of altruism is not suitable to explain behavior in experimental dictator-type games.[12] For example, it cannot account for subjects who maximize only their own monetary value regardless of the behavior of the others. Experiments have shown that such subjects, often referred to as selfish, constitute a large proportion of all subjects [100]. It would be most appropriate to model such players with $\lambda = 0$ and $\gamma = 0$. Indeed, the choices of 38 percent of our subjects are consistent with $\lambda = 0$ and $\gamma = 0$.

[12] This is acknowledged by Levine [132].

Coefficients of Inequity Aversion

Assume that the preferences of subject i comply with utility function (5.8) for some (unknown) coefficients α_i and β_i. The preferences of subject i over the angles in the circle-test are then described by the utility function

$$u_i(\tau_i) = \cos\tau_i - \alpha_i \max\{\sin\tau_i - \cos\tau_i, 0\} - \beta_i \max\{\cos\tau_i - \sin\tau_i, 0\} \ . \quad (5.13)$$

Proposition A.2, given in Appendix A, asserts that for all feasible coefficients α_i and β_i this utility is maximized only if $\tau_i \in [0, \frac{\pi}{4}]$. In this interval the own payoff ($\cos\tau_i$) always exceeds the payoff of the other ($\sin\tau_i$). Consequently,

$$u_i(\tau_i) = \cos\tau_i - \beta_i(\cos\tau_i - \sin\tau_i) \ ,$$

which implies that the coefficient α_i does not affect player i's choice of the angle in the circle-test.

A selfish player with $\beta_i = 0$ maximizes her utility by choosing $\tau_i = 0$. Any player with $\beta_i \geq 0.5$ maximizes her utility by choosing $\tau_i = \frac{\pi}{4}$. If a player maximizes her utility by an angle $\tau_i^* \in (0, \frac{\pi}{4})$ then[13]

$$\beta_i = \frac{\sin\tau_i^*}{\cos\tau_i^* + \sin\tau_i^*} \ . \quad (5.14)$$

There is a one-to-one correspondence between angles in the interval $[0, \frac{\pi}{4})$ and coefficients β in the interval $[0, 0.5)$. Using (5.14) we estimate the β-coefficients of our subjects. The distribution of estimated coefficients is shown in Fig. 5.14. Our distribution puts larger weight on low β-coefficients, compared to distribution (5.9). Over 35 percent of our subjects appear selfish, with $\beta_i = 0$. Further 24 percent of our subject's estimated β-coefficients are below 0.3. Only 10 percent of the estimated coefficients is in the interval $[0.5, 1)$, which is much less than the estimate of Fehr and Schmidt that β-coefficients of 40 percent of their subjects are 0.6. However, we cannot reject that $\beta_i > 0.71$ for these 10 percent of our subjects, which would imply that they cooperate in the Nash equilibrium of our experimental games. About 5 percent of decisions cannot be explained by the model (5.13).

[13] See Proposition A.2 in Appendix A to this chapter.

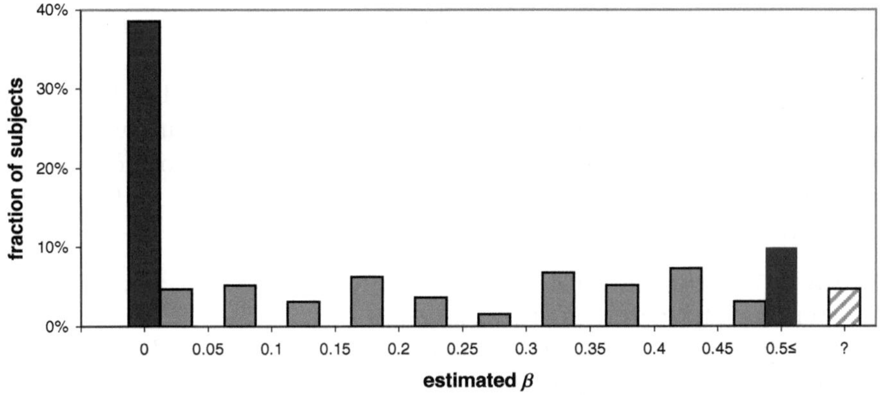

Fig. 5.14. The distribution of estimated β-coefficients of inequity aversion among our experimental subjects. Dark colored column "0" represents the proportion of subjects with $\beta_i = 0$. Dark colored column "0.5≤" represents the proportion of subjects with $0.5 \leq \beta_i < 1$. The patterned column "?" represents the proportion of subjects whose choice can not be explained as maximization of utility function (5.13). The remaining columns represent the proportions of subjects whose estimated coefficient is between the corresponding two values

5.6.5 Summary

The experimental games Γ^I (treatment I), Γ^{II} (treatment II), and Γ^{III} (treatment III) , were not played as suggested by the theory in Chap. 4. If monetary payoffs in experimental games, as assumed in Chap. 4, correctly describe preferences of rational players, and if this is common knowledge, then we should observe no cooperation. This is not what we see in our experiments: in each of the treatments we observe, on average, substantial cooperation. Furthermore, in games with endogenous network formation, cooperation was sustained at a high level during all but the last few periods. As described below, in each treatment we observe positive average levels of cooperation even in the very last period.

The inconsistency between the theoretical predictions and the experimental evidence is not surprising in view of many accounts of similar discrepancies for public goods games and other social dilemma games. It is frequently suggested that we may need to model our subjects different than as rational own *monetary earning* maximizers. In this section we therefore study whether behavior in our experiments can be explained by assuming that our subjects are rational *utility* maximizers, but that their utility functions depend not just on their own monetary earnings but on monetary earnings of all players in their reference group.

Such subjects are called other-regarding. Several models have been proposed in the literature to account for heterogeneity among the subjects in their regard for others. In this section we study two such models, one put forward by Levine [132] and one by Fehr and Schmidt [95]. Our conclusions can be summarized as follows.

Result 5.7. COOPERATION IN EQUILIBRIA OF GAMES Γ^{I}, Γ^{II} AND Γ^{III} WITH OTHER-REGARDING PLAYERS
If the coefficients are distributed as described by the models of altruism and inequity aversion then there is no cooperative play in any Nash equilibrium.

The above result follows from Propositions 5.1 and 5.2. It is interesting to contrast the theoretical prediction, that in the last period of the experimental game no player with an established link will cooperate, with experimental evidence. The ratio of subjects that establish a link and cooperate in the last period is 11% in treatment I, 30% in treatment II and 7% in treatment III. In our view the inconsistency between the prediction and the last period behavior is substantial, in particular in treatment II.

In order to estimate the other-regarding preferences of our own subjects, given by their social value orientation, we ran a circle-test before each experimental session. We can couple the decision of each subject in the circle-test with her behavior in the last period of the subsequent experimental game. This permits us to study the relation between our subjects' regard for others and the last-period cooperative play. Proposition 5.1 asserts that, in a Nash equilibrium, condition (5.6) must be satisfied for each player that cooperates and establishes at least one link. In regard to inequity aversion, Proposition 5.2 asserts that $\beta_i \geq \frac{n-1}{n+1}$ (> 0.7 for $n = 6$) must be satisfied for each such player in a Nash equilibrium. The experimental evidence, however, is not consistent with these predictions. Among the subjects that in the last period establish at least one link and cooperate, there are 52 percent whose coefficients of altruism violate (5.6) and 76 percent whose coefficient of inequity aversion is below 0.5.

Result 5.8. SOCIAL PREFERENCES OF OUR SUBJECTS
The observed last period cooperation in our treatments cannot be consistently explained by the social preferences of our subjects.

In Chap. 6 we explore a model of behavioral dynamics in a network dilemma game Γ^{I} when played by boundedly rational, and possibly other-regarding, players. We report there some success in capturing the actual dynamics in our experimental games.

5.7 Discussion and Conclusion

In this chapter we have presented an experimental investigation of the dynamics of behavior in repeated network dilemma games. We implement an environment with mutual link formation and a low outside option. We report on two experimental treatments, one imposing a strict linking constraint and another with unconstrained linking. We also compare the dynamics of behavior in experimental treatments with the dynamics of behavior in a control treatment where the network structure is exogenously imposed. The theoretical equilibrium analysis permits exclusion and cooperation only in the treatment with linking constraints.

The following are our main conclusions: (i) Free choice of social links drastically increases cooperation. In both experimental treatments cooperation is significantly higher compared to the control treatment. Cooperation levels in the experimental treatments appear to be constant or even increasing. (ii) Contrary to the subgame perfect predictions, subjects use the costly option to exclude those who tend to defect, even in absence of linking constraints. In all experimental treatments cooperators that refuse linking to defectors account for most of social exclusion. (iii) Cooperation does not appear to be affected by linking constraints. Average levels of cooperation are similarly high (around 70 percent) in both experimental treatments. (iv) Linking constraints increase segregation between cooperators and defectors. In treatment I without linking constraints a link between a cooperative and a defective subject is established in 67 percent of possible cases. In treatment II with strict linking constraints such a link is established in only 10 percent of possible cases. (v) Other-regarding preferences of our subjects cannot explain the levels of cooperation observed in the last-period. In addition, the equilibrium analysis of our experimental games using models of altruism and inequity aversion agree with our standard equilibrium prediction and fail to explain the observed last-period behavior.

6

Dynamics of Partner Choice and Cooperation

Ever since Robert Axelrod initiated his computer tournament aimed at finding the winning strategy in a repeated prisoner's dilemma game have simulations and tournaments been the preferred methodology of social scientists investigating the phenomena of reciprocity and cooperation. Simulations are attractive because they can often be applied when theoretical deduction is not possible. Even simple models of social interaction often produce complex dynamics that are hard to characterize with mathematical analysis. This is especially likely when the system under study has many elements that interact with each other in many different ways. The dynamics of such a system usually depends on the emerging interaction pattern as well as on the private experiences of each agent. Simulations may yield paths of system dynamics given by any set of behavioral rules and thus provide an insight into effects of any particular modeling assumption. Approaching the same task mathematically is often much more difficult.

In this chapter we use computer simulations to study the dynamics of partner choice among agents that follow simple behavioral rules in a stylized model of a social dilemma embedded in an endogenous network. We identify the conditions under which a norm can emerge that prescribes individual exclusion of players that do not cooperate, and the consequences of this norm for social cooperation. The modeling variables are the inclination to fairness among the players and their comprehension of the behavioral strategies used by other players. Even though our model considers only basic behavioral rules it produces rich dynamics that vary to a great extent across distinct simulation conditions.

In the following section we discuss how an evolutionary view to behavior may give an important alternative approach to that of modeling rational agents in economics. We then shortly describe our dynamic model and discuss some related approaches in the literature, and our contribution. The model is formally introduced and analyzed in Sec. 6.2. Medium and long-run simulations are described in detail in Sec. 6.4, and their application is demonstrated for the case of selfish players. In Sec. 6.5 we use simulations to study the dynamics of behavior among non-selfish players. In Sec. 6.6 we discuss the results of our simulations and some possible extensions of our dynamic model. We briefly summarize the chapter in Sec. 6.7.

6.1 Introduction

In his seminal tournament Axelrod [94] asked a selection of researchers to submit strategies that would compete against each other in a finitely repeated prisoner's dilemma game. Even though many sophisticated strategies were submitted the winning one was simple. Called tit-for-tat, it was characterized by cooperating in the first period and subsequently repeating the action that the opponent played in the previous period. The strengths of this strategy were unconditional reciprocation and simplicity, so that other strategies were quickly able to learn to cooperate with it. Reciprocity and simplicity also characterized the winner of a computer simulation of the evolution of strategies in which strategies were not submitted but randomly generated by the computer.

Two kinds of evolutionary approaches are most prominent among the numerous simulation studies that have since been implemented by social scientists, computer scientists, mathematicians, economists and others. The first approach is to consider a limited set of strategies, usually a convex hull of those that describe intuitive modes of behavior. The dynamics of proportions of agents playing different strategies is then analyzed in a simulated environment under the assumption that the change in proportion of each strategy is a function of its relative success. In particular, the use of successful strategies is assumed to grow and of unsuccessful ones to decline, although they never quite disappear. This assumption can be motivated by conjecturing that successful strategies are imitated more often than unsuccessful ones, or that agents playing successful strategies have more offspring than those playing unsuccessful ones. See [146, 147] for examples of this approach.

The second approach is to set complexity bounds on prospective strategies and specify the rules by which new strategies evolve from previous ones. Strategies may, for example, be structured as cellular automata in which case complexity can be bounded by limiting the maximal number of automata states. Only a small subset of strategies is present in the population at any certain time but new strategies may appear by way of random mutations of the most successful past strategies. This potentially permits a very large set of prospective strategies that is not limited by the intuition of the researcher. Evolutionary simulations of this kind may yield unexpected strategies and are useful exploratory studies but may be difficult to interpret. A number of recent papers using various evolutionary approaches for the study of the emergence of cooperation are surveyed in [148].

The main reason to adopt computer simulations as a methodology of research in economics is to facilitate the study of models that reach beyond the traditional assumption of fully rational, informed, computationally capable and selfish economic agents. This assumption gives rise to accurate predictions about aggregate market behavior [149] but it offers an oversimplified view of human behavior and its predictions at the micro-level often dramatically fail. One example is its failure to explain high levels of cooperation in experimental finitely repeated two-player prisoner's dilemma games. Many other examples can be found in laboratory experiments inspired by game theory [119]. Empirical evidence as well as casual experience yield further phenomena that the standard theory cannot explain, such as altruism, reciprocity, and conformism with social norms [150].

The postulates about rationality, perfect foresight and selfishness are contested by the models of bounded rationality. Although people may try to act rationally the requirements of information and computational abilities that standard models assume are rarely met [151]. Customers have strong feelings about the perceived fairness of firms and worker's perception of what constitutes a fair wage constrains the wage setting of firms [99, 152]. It is striking that when people have incomplete information or use behavioral rules that are constrained in complexity they may reach far higher levels of cooperation than suggested by standard theory [14, 75].

A common approach to modeling bounded rationality is to consider agents that follow simple behavioral rules. By way of learning and adaptation they occasionally change their heuristics. The updating dynamics may lead to a stable constellation of rules in the population of such agents. Rather than being the consequence of rational deliberation the emerging behavior constitutes the outcome of adaptation of agents to a complex environment. Using this approach one can contribute to an explanation of reciprocity, conformism with social norms, and cooperation. Because of analytical constraints these models are often analyzed with the help of computer simulations.

We introduce in this chapter such a model to study conditions under which a norm that prescribes exclusion of defective players can survive, and the consequences of this norm for the dynamics of cooperation. Social exclusion of defectors may be seen as a form of punishment. It may, however, have additional effects on the evolution of behavior, especially because it affects the social structure and the pattern of future interactions. We therefore discuss our approach in relation to the literature on the evolution of norms and punishment behavior as well as the literature on spatial evolution of cooperation.

That cooperation may be supported by reciprocation was already demonstrated by the success of the tit-for-tat strategy in Axelrod's tournament (although some evolutionary support for this conjecture already existed in the biological literature, see e.g. [153]). This strategy, however, is much less successful in multi-player games and some kind of retaliation that is directed solely at non-cooperators is needed to sustain cooperation. Boyd and Richerson [154, 155] refer to such directed retaliation as retribution and study its evolutionary success. Retribution, if costly, can be seen as an altruistic act. Punishment is beneficial to the group, but costly to the individual, and selection favors individuals that do not punish. This is commonly referred to as the "second order" free riding problem [156]. Several solutions to this problem have been suggested. One may, for example, consider punishers who punish any order free riders, thus following a metanorm [157], cooperators who always retaliate [158], higher order punishments [159], and inter-group competition [160].

The structure of social interaction, on the other hand, may also crucially affect the evolution of cooperation. This has first been demonstrated by Nowak and May [17]. Large number of models of adaptive play on different interaction structures, in what became known as models of local interaction, have since been analyzed using simulations [161, 162], tournaments [163] and theoretical analysis [18, 164]. A particularly thorough investigation is reported in [165], using simulations to compare dynamics of cooperation across different structures of interaction, adaptive processes and strategy spaces.

A common assumption in models of the evolution of retaliation and local interaction is that successful strategies grow by way of imitation or conformism. In particular, best response dynamics are not considered. Eshel, Samuelson and Shaked [18] claim that "altruism has no hope in the world of best responders", as defection is the unique best response. This, however, is true only in the world of myopic best responders, that is, players who look for strategies that maximize only their immediate payoff. We demonstrate in this chapter that cooperation may be sustained under best response dynamics when players understand the reactions of other players and take them into account when calculating their payoff.

Players that are sufficiently sophisticated to include their knowledge about strategies used by other players when calculating their best response are usually referred to as level-2 players [166], as opposed to level-1 players who play best response to previous period actions and level-0 players who play randomly. That many people exhibit what is called strategic sophistication was demonstrated in experiments [167, 168]. Between 45% and 53% of experimental subjects were classified as level-2 players. Such players may have significant impact on the dynamics of play, but have so far been largely neglected in dynamic models of adaptive behavior. Only the simplest modes of behavior, such as imitation or myopic best response, or most sophisticated ones, such as perfect foresight, are usually considered. We propose in this chapter a simple way of modeling level-2 players via limited forward looking.

Players with forward looking of limited scope $r \geq 0$ choose strategies that maximize their utility across r periods. Players with $r = 2$ consider reactions of other players to their own behavior in their utility calculations in accordance with level-2 sophistication. Players with $r = 1$ choose myopic best response in accordance with level-1 sophistication. Players with $r = 0$ do not maximize their future payoff but make random actions, in accordance with level-0 sophistication. To study the effect of increased forward looking we also study players with $r = 4$, although not with level-4 sophistication. A reference to a few other models of limited foresight, with a short discussion of their shortcomings, is given in [169]. Jehiel [170] proposes an intricate notion of an equilibrium with limited foresight, which we discuss in Sec. 6.6.2.

Consideration of best response dynamics permits us to make another contribution, by modeling the dynamics of behavior among players with non-selfish preferences. That people may have sentiments for the well-being of other people, and may not necessarily be selfish, has been debated by philosophers as well as economists (a historical reference is Adam Smith [171], see also [172]). It has been observed empirically and demonstrated via laboratory experiments (see e.g. the dictator game experiments in [135]). Recently a number of approaches to model non-selfish players have been suggested [128, 95, 132, 131]. In this chapter we consider two simple models from this literature, the model of altruism by Levine [132] and the model of inequality aversion by Fehr and Schmidt [95], and study the dynamics of norms and cooperation among altruistic or inequity averse players. Because these models require comparisons of prospective payoffs they are incompatible with assumptions of imitation or conformism made in other evolutionary studies. In our model, in which we assume best-response behavior, they can be implemented in a straightforward manner.

Best response has previously been used in the framework of the dynamics of conventions. Ellison [173] and Blume [174] both studied the evolution of coordination among myopic best responders embedded in a local interaction network. They noted that non-perturbed dynamics may converge to several equilibria, but if the process is perturbed by small errors the risk dominant equilibrium is most likely to be selected. Their approach is based on observations made in [175] and [176] that the dynamics in games with noise may drastically differ from that in games without noise. Small perturbations may result in a process that, in the long run, spends considerably more time in some states than in the others. In this sense we may say that the dynamics select among the equilibria in the long run. We adopt the approach taken in [176] to detect the long-run equilibria in our model.

This is how we set up our model. We assume that a relatively small group of players is playing the prisoner's dilemma game embedded in an endogenous network. In each period players simultaneously choose with whom to play and what action to play. Each pair of players play the prisoner's dilemma game when both mutually chose each other, and earn the outside option otherwise. Our aim is to study the dynamics of the norm that prescribes cooperation and exclusion of players that defected in the previous period. We conjecture that full cooperation will emerge if sufficient number of players follows such a norm. Whether such a situation can be sustained in the long run may depend on forward looking and social preferences of the players as well as on the outside option value. We run computer simulation across different scopes of forward looking, different outside options, and different constellations of social preferences in the group. Our main conclusions are that (i) sufficient forward looking is as important for sustaining cooperation as is sufficient regard for others, (ii) cooperation may be sustained among sufficiently forward looking selfish players, and (iii) even though assuming other-regarding preferences increases the set of cooperative equilibria, there are no qualitative differences in the long-run dynamics between groups of other-regarding and groups of selfish players. Cooperation increases considerably only when extremely high altruism or inequality aversion is assumed.

A few other studies exist that discuss the evolution of cooperation when one's interaction partners may be endogenously chosen or refused. The replicator dynamics within a small set of simple strategies in repeated two person prisoner's dilemma game with options of exit or partner change are explored in [49, 47, 177, 178]. The evolution of altruism in Public Goods games with exit possibility is investigated in [179] and [180]. The evolutionary success of optimistic cooperators who often play the game, facing pessimistic defectors who are more likely to take the outside option is investigated in [181] and [182]. Several models of random matching according to endogenously chosen probability distributions are developed in [183, 164, 72]. A model of network formation with exogenously defined linking rules is described in [184]. The effects of risk preferences on migration and cooperation in a model with heterogeneous players are studied in [185]. The success of various strategies in a finitely repeated prisoner's dilemma with endogenous matching is explored through computer tournaments [186, 187]. Genetic algorithms are used to study the evolution of cellular automata playing prisoner's dilemma in a partner choice and refusal environment [188, 48].

All of these studies assume that behavior spreads via imitation or a related adaptive model and that players are constrained in the number of partners they may choose. In Chap. 4 we show that if linking is constrained then cooperation can be achieved in subgame-perfect equilibria of finitely repeated network dilemma games even among rational and selfish players with perfect foresight. In contrast, we study in this chapter a repeated network dilemma game with costless and unconstrained linking. Our intuition that boundedly rational players may achieve cooperation in such setting is confirmed in our dynamic model.

6.2 Adaptive Play

In this chapter we consider network dilemma games with unconstrained and costless linking, introduced in Chap. 3. We refer to such games as *basic games*.

Given a basic game $\Gamma = \langle N, J, \pi \rangle$ with n players, the following dynamic model is considered. Let $t = 0, 1, 2, ...$ denote successive time periods. The basic game is played once each period. Let $a_i(t)$ be the action and let $p_i(t)$ be the linking choice of player i at time t. The profile of moves at time t is $(a(t), p(t)) = ((a_1(t), .., a_n(t)), (p_1(t), ..., p_n(t))) \in J$. The *history* of moves up to time t is the sequence $h(t) = (a(0), p(0), a(1), p(1), ..., a(t), p(t))$. The set $H[t] = J^t$ consists of all possible time t histories.

At each time t player i is characterized by a *behavioral strategy* (or simply: strategy), defined as a mapping

$$s_i[t] : H[t-1] \to A_i \times P_i$$

assigning a move $(a_i(t), p_i(t)) = s_i[t](h(t-1))$ to each possible history $h(t-1)$. For convenience we omit the time parameter $[t]$ whenever we refer to generic moves, strategies, and histories. We denote by $p_i(s_i \mid h)$ the linking choices and by $a_i(s_i \mid h)$ the action chosen by the strategy s_i given history h.

We assume that players exhibit a fair amount of inertia in choice of their strategies. With high probability they keep the same strategy across several periods. Only occasionally a player realizes what strategies are played by the other players and updates her own strategy by choosing one that maximizes her utility given the strategies of the other players. An updating player assumes that the other players will continue to play the same strategies for a few more periods, and that she herself will be playing her new strategy for a few following periods. These assumptions may be viewed as a boundedly rational behavior of the players facing the difficult task of predicting the future behavior of the other players.

When maximizing her future payoffs the player may consider only the payoffs in the coming period, in which case we say that player chooses a *myopic best response*. A standard result is that defection is the unique myopic best response in a multi player prisoner's dilemma game. We show here that in our basic game a player following the myopic best response never cooperates over an established link. The aim of this chapter is to extend the analysis by considering players which, to a limited extent, understand the future consequences of their actions. This is modeled by assuming that players behave in order to maximize their payoff across several periods of play. We discuss when cooperation among players with such "limited forward looking" can be sustained.

6.2.1 Strategy Set

Below we introduce the set of strategies considered in the simulations. Rather than allowing for all feasible strategies at each time t we consider a fixed set of possible behavioral strategies. The strategies are independent of the time period. This is because we assume that players have a limited memory and consider only the most recent part of history, even though the length of history increases with time. In particular, we assume for simplicity that the strategies are independent of all past periods except the most recent one. We also assume that strategies depend only on past actions, but are independent of past linking choices.

Further, we assume action consistency and linking consistency. Action consistency requires that strategies are either fully cooperative or fully defective. That is, a player cooperates in each period she is playing a cooperative strategy and defects in each period she is playing a defective strategy. Linking consistency requires that a player proposes a link to another player playing strategy s if and only if she also proposes links with all other players playing strategy s. We do not consider mixed strategies.

The resulting set consists of four intuitively plausible behavioral strategies. The benefit of a small set of feasible strategies is the tractability and clarity of interpretation of the simulation dynamics.

We assume that each player is of one of the following types:

(CA) (Non-exclusive) cooperator. If player i is of this type she proposes links to all other players and cooperates in the prisoner's dilemma game. For any time t the corresponding strategy s_i^{CA} is defined by

$$s_i^{CA}(h(t-1)) = (C, p_i^{CA}) \ ,$$

and the vector of proposed links p_i^{CA} by

$$p_{ij}^{CA} = \begin{cases} 1 & \text{if } j \neq i \ , \\ 0 & \text{if } j = i \ . \end{cases}$$

(CE) Exclusive cooperator. If player i is of this type she proposes links only to those other players that cooperated in the previous period, and cooperates in the prisoner's dilemma game. For any time t the corresponding strategy s_i^{CE} is defined by

$$s_i^{CE}(h(t-1)) = (C, p_i^{CE}) \ ,$$

and the vector of proposed links p_i^{CE} by

$$p_{ij}^{CE} = \begin{cases} 1 & \text{if } a_j(t-1) = C \ , \\ 0 & \text{if } a_j(t-1) = D \text{ or } i = j \ . \end{cases}$$

(DA) (Non-exclusive) defector. If player i is of this type she proposes links to all other players and defects in the prisoner's dilemma game. For any time t the corresponding strategy s_i^{DA} is defined by

$$s_i^{DA}(h(t-1)) = (D, p_i^{DA}) ,$$

and the vector of proposed links p_i^{DA} by

$$p_{ij}^{DA} = \begin{cases} 1 & \text{if } j \neq i , \\ 0 & \text{if } j = i . \end{cases}$$

(DE) Exclusive defector. If player i is of this type she proposes links only to those other players that cooperated in the past period, but herself defects in the prisoner's dilemma game. For any time t the corresponding strategy s_i^{DE} is defined by

$$s_i^{DE}(h(t-1)) = (D, p_i^{DE}) ,$$

and the vector of proposed links p_i^{DE} by

$$p_{ij}^{DE} = \begin{cases} 1 & \text{if } a_j(t-1) = C , \\ 0 & \text{if } a_j(t-1) = D \text{ or } i = j . \end{cases}$$

We refer to a player of type CA or CE as *cooperative*, and to a player of type DA or DE as *defective*. We refer to a player of type CE or DE as *exclusive*, and to a player of type CA or DA as *non-exclusive*. In Table 6.1 we use the same nominations to classify the corresponding strategies. The non-exclusive strategies are independent of the history, while the exclusive strategies depend on the most recent actions only. Denote $S_i = \{s_i^{CA}, s_i^{CE}, s_i^{DA}, s_i^{DE}\}$ and let $S = \{(s_1, ..., s_n) \mid s_i \in S_i\}$ be the space of strategy vectors. Let $\mathcal{T} = \{CA, CE, DA, DE\}$ denote the set of available types.

We are interested in the long run dynamics of norms prescribing exclusion of defectors, and the impact of the exclusion in the long run success of cooperation. The four strategies defined above represent the modal types of behavior for our analysis. A notable omission is the absence of tit-for-tat type strategies, that is, strategies which prescribe cooperation as long as certain number of other players cooperate, and defection otherwise. In n-player prisoner's dilemma games defection of a single player negatively affects all other players, those who defect as well as those who cooperate. Even if the intent of defection is to punish a subgroup of defectors it may induce only more defections by the remaining cooperators, and the play may slide into the uniform defection. Exclusion of defectors, on the other hand, is a form of retaliation

Table 6.1. Classification of strategies according to cooperation and exclusiveness

	always link	exclude defectors
cooperate	$\mathbf{s}^{CA}$	$\mathbf{s}^{CE}$
defect	$\mathbf{s}^{DA}$	$\mathbf{s}^{DE}$

and does not risk being misinterpreted for defection by the remaining players. Rather than tit-for-tat strategy, we study in this chapter the success of the *out-for-tat* strategy s_i^{CE}.

6.2.2 Updating Rule

We describe next the strategy updating rule. Players use the same rule in all periods in which they update their strategy. For convenience we omit the time index $[t]$ whenever it is not essential. Let a be the vector of actions at time $t-1$ and let $\mathbf{s} = (s_1, ..., s_n) \in S$ be the vector of strategies that players use at time t. The profile of moves at time t is then determined by $\mathbf{s}$ and a. Let $a_i(s_i) \equiv a_i(s_i \mid h(t-1))$ be the action, and let $p_i(s_i \mid a) \equiv p_i(s_i \mid h(t-1))$ be the linking choice induced at time t by strategy s_i, given actions a at time $t-1$. Let $a_{-i}(\mathbf{s}) = (a_1(s_1), ..., a_{i-1}(s_{i-1}), a_{i+1}(s_{i+1}), ..., a_n(s_n))$ be the vector of actions and let $p_{-i}(\mathbf{s} \mid a) = (p_1(s_1 \mid a), ..., p_{i-1}(s_{i-1} \mid a), p_{i+1}(s_{i+1} \mid a), ..., p_n(s_n \mid a))$ be the vector of linking choices of players N/i induced by $\mathbf{s}$. The vector of strategies that players use at time t is denoted by $\mathbf{s}[t] = (s_1[t], ..., s_n[t])$.

Players begin with an initial vector of strategies $\mathbf{s}[1] \in S$ at time $t = 1$.[1] At each time $t \geq 2$, before making her move, player i with some fixed probability $q_i \in (0, 1)$ updates her own strategy, by observing the vector of strategies at time $t-1$ and maximizing her payoff across the next r_i periods, under the assumption that the strategies of the other players do not change in that time span. Formally, let $\mathbf{s} = (s_1, ..., s_n) \in S$ be the vector of strategies at time $t-1$, and let

$$\mathbf{s}' = (s_1, ..., s_{i-1}, s_i', s_{i+1}, ..., s_n)$$

be the vector of strategies during time periods $[t, t+r_i]$ assuming that player i changes her strategy from s_i to s_i' at t, keeps it during $[t, t+r_i]$, and that players N/i do not change their strategies during $[t-1, t+r_i]$.

[1] In our simulations we will consider different initial strategy profiles. We specify them later.

Under these assumptions, the payoff of player i across the r_i periods following her update is

$$x_i(s'_i \mid \mathbf{s}) = \pi_i(a_i(s'_i), p_i(s'_i \mid a(\mathbf{s})), a_{-i}(\mathbf{s}), p_{-i}(\mathbf{s} \mid a(\mathbf{s}))) \\ + (r_i - 1) \cdot \pi_i(a_i(s'_i), p_i(s'_i \mid a(\mathbf{s}')), a_{-i}(\mathbf{s}'), p_{-i}(\mathbf{s}' \mid a(\mathbf{s}'))) \; . \quad (6.1)$$

The first row of (6.1) is the payoff player i receives at time t, that is, when the remaining players believe she is playing the old strategy $s_i = s_i[t-1]$, but in fact she is playing the new strategy s'_i. The second row of (6.1) is the payoff player i receives during the subsequent $r_i - 1$ time periods, that is, when the remaining players realize she is playing the new strategy s'_i. This calculation is correct if no updates take place during $[t-1, t+r_i]$, aside from the initial update by player i.

If player i is updating at time t she chooses a strategy $s_i[t]$ such that

$$s_i[t] \in \arg\max_{s'_i \in S_i} x_i(s'_i \mid \mathbf{s}[t-1]) \; . \quad (6.2)$$

If this payoff is maximized by the current strategy of the player, the player keeps her current strategy. If not, she randomly selects one of the optimal strategies. The parameter r_i is called the *scope of forward looking* of player i. If player i does not update her strategy, which happens with probability $(1-q_i)$, we say that she exhibits 'inertia', i.e. she maintains her current strategy.

The assumption that strategies of other players can be observed is not very restrictive. To see this, note that (i) actions of all players are always observable and (ii) whenever at least two players defect the linking behavior of all players in the next period reveals their type.[2] This means that whenever at least two players defect strategies of all players are revealed in the following period. For simplicity we assume that a player observes all strategies whenever she is updating her strategy.

6.3 The Basic Process

When at some time $t \geq 2$ a player updates her strategy her choice depends only on the vector of strategies played a time $t-1$.[3] The probability that players will play strategies $\mathbf{s}[t]$ at time t depends therefore

[2] If no player defects then in the following period all players propose all links and do not reveal whether they are exclusive or not. If only one player defects then this player, regardless of her type, proposes all links in the following period.

[3] We assume that time starts with $t = 1$, but the first update may take place only after the first round, that is at $t = 2$. This is because players need to observe at least one period of play before they can update their strategies.

only on the vector of strategies $\mathbf{s}[t-1]$ played at time $t-1$, and not on the strategies played during the previous time periods. Furthermore this probability is stationary, that is, it is independent of t. The updating rule outlined above thus generates a Markov chain with the finite state space S. This Markov chain is completely characterized by its transition matrix M in which for each $\mathbf{s}, \mathbf{s}' \in S$ the entry

$$M_{\mathbf{ss}'} = prob[\mathbf{s}[t] = \mathbf{s}' \mid \mathbf{s}[t-1] = \mathbf{s}] \tag{6.3}$$

gives the probability that the process is in state $\mathbf{s}'$ at time t, given that it was in state $\mathbf{s}$ at time $t-1$.[4] The value of $M_{\mathbf{ss}'}$ is also called the probability of *transition* from state $\mathbf{s}$ into state $\mathbf{s}'$. The value of $M_{\mathbf{ss}}$ is the probability that the process will not make any transition from state $\mathbf{s}$. For each state $\mathbf{s}$ the corresponding row of the matrix represents the probability distribution over transitions from state $\mathbf{s}$, and satisfies

$$\sum_{\mathbf{s}' \in S} M_{\mathbf{ss}'} = 1 \ . \tag{6.4}$$

We refer to this particular discrete-time Markov process as the *basic process*.

6.3.1 Absorbing States of the Basic Process

Let $\mathbf{s} = (s_1, ..., s_n)$ be an absorbing state. By definition the set $\{\mathbf{s}\}$ is recurrent, which implies that $M^{\tau}_{\mathbf{ss}'} = 0$ for every state $\mathbf{s}' \neq \mathbf{s}$ and all $\tau \in \mathbb{N}$. It then follows from (6.4) that $M_{\mathbf{ss}} = 1$, from (6.3) that $\mathbf{s}[t] = \mathbf{s}$ implies $\mathbf{s}[t+1] = \mathbf{s}$ and, indeed, that $\mathbf{s}[t+\tau] = \mathbf{s}$ for all $\tau \in \mathbb{N}$. This may hold only when

$$s_i \in \arg\max_{s'_i \in S_i} x(s'_i \mid \mathbf{s}), \tag{6.5}$$

for each player i. It is easy to verify that the opposite argument is also true, which proves the following Proposition.

Proposition 6.1. *State* $\mathbf{s}$ *is absorbing if and only if (6.5) holds for each player* i.

[4] The definition (6.3) requires $t \geq 2$. Because of the stationarity of the process this probability is well defined, that is, it is independent of t.

In other words, in an absorbing state, each player is playing a strategy that is a best response with limited forward looking to the strategies played by the other players. In this sense we may say that the process is in equilibrium when it reaches an absorbing state.

In each state the set of possible transitions depends only on the numbers of players of each type. This is shown in the proof of Proposition 6.2, below. An intuitive explanation is that the strategies do not discriminate among the players on the basis of their identities. If players j and k play the same strategy at time $t-1$, then each player $i \neq j,k$, using one of the strategies outlined above, will react to players j and k in the same way at time t. Formally, for any type $T \in \mathcal{T}$ let

$$n^T(\mathbf{s}) = \left|\{i \in N \mid s_i = s_i^T\}\right|$$

be the number of players of type T given the vector of strategies $\mathbf{s}$.[5] For each $\mathbf{s} \in S$ we refer to the 4-tuple

$$z(\mathbf{s}) = (n^{CA}(\mathbf{s}), n^{CE}(\mathbf{s}), n^{DA}(\mathbf{s}), n^{DE}(\mathbf{s}))$$

as the *type distribution* in state $\mathbf{s}$. Let

$$Z = \{(n^{CA}, n^{CE}, n^{DA}, n^{DE}) \mid \\ n^{CA} + n^{CE} + n^{DA} + n^{DE} = n,\ n^T \geq 0 \text{ for each } T\}$$

be a set of possible type distributions. For each type distribution $z \in Z$ let

$$S_z = \{\mathbf{s} \in S \mid z(\mathbf{s}) = z\}$$

be the set of all states with type distribution z. Sets S_z partition S into groups of states with equal type distribution.

Let us introduce a short notation for the states in which all players play the same strategy. For a type $T \in \mathcal{T}$ let $\mathbf{s}^T = (s_1^T, ..., s_n^T)$ be the state in which each player i plays the strategy s_i^T. The corresponding type distributions are

$$\begin{aligned} z(\mathbf{s}^{CA}) &= (n, 0, 0, 0), \\ z(\mathbf{s}^{CE}) &= (0, n, 0, 0), \\ z(\mathbf{s}^{DA}) &= (0, 0, n, 0), \\ z(\mathbf{s}^{DE}) &= (0, 0, 0, n). \end{aligned}$$

The following proposition states that for each $z \in Z$ either all or none of the states with type distribution z are absorbing. The proof is given in Appendix A.

---

[5] Cardinality $|A|$ gives the number of elements in the set A.

Proposition 6.2. *Let $z \in Z$ be a type distribution. State $\mathbf{s} \in S_z$ is absorbing if and only if each of the states in S_z is absorbing. State $\mathbf{s} \in S_z$ is transient if and only if each of the states in S_z is transient.*

Propositions 6.1 and 6.2 facilitate characterization of absorbing states of the basic process. To prove that states with type distribution z are absorbing it suffices to show that in some state $\mathbf{s} \in S_z$ strategies of all players satisfy condition (6.5). To prove that states with type distribution z are not absorbing it suffices to show that the strategy of some player in some state $\mathbf{s} \in S_z$ violates condition (6.5), that is, her strategy is not a best response with limited forward looking to strategies played by the other players.

A non-absorbing state is either transient or belongs to a recurrent set. If the absorbing states are known, characterization of recurrent sets reduces to the characterization of the set of transient states. For this we use the following result.

Proposition 6.3. *A state $\mathbf{s}$ is transient if some transient or absorbing state is accessible from $\mathbf{s}$. In particular, a state $\mathbf{s}$ is transient if $M_{\mathbf{ss}'} > 0$ for some transient or absorbing state $\mathbf{s}' \neq \mathbf{s}$.*

Proof. By definition, state $\mathbf{s}$ belongs to a recurrent set if all states accessible from $\mathbf{s}$ also belong to the same recurrent set. Also by definition, a transient or absorbing state never belongs to a (non-singleton) recurrent set. If a transient or absorbing state is accessible from $\mathbf{s}$, this implies that $\mathbf{s}$ does not belong to any recurrent set, and is therefore transient. □

We implement the assertion of Proposition 6.3 to detect the transient states of the basic process via the following algorithm.

Algorithm 1

1. Calculate the transition matrix M.
2. Determine the absorbing states S^A of the process.
3. Let S^0 be the set of states $\mathbf{s} \in S^0$ such that $M_{\mathbf{ss}'} > 0$ for some absorbing state $\mathbf{s}' \neq \mathbf{s}$.
4. Repeat for $\tau = 1, 2, ...$:
   let S^τ be the set of states $\mathbf{s} \in S^\tau$ such that $M_{\mathbf{ss}'} > 0$
   for some $\mathbf{s}' \in S^{\tau-1}$,
   until $S^\tau = S^{\tau-1}$.

Proposition 6.4. *Let, for $\tau = 0, 1, ..., \overline{\tau}$, the sets S^τ be yielded by the Algorithm 1. The process has no recurrent sets, aside from the absorbing states, if and only if $S^0 \cup S^{\overline{\tau}} = S$.*

Proof. An iterative application of Proposition 6.3 implies that for each τ all states in S^τ are transient. By the definition of the updating process, regardless of the state, the event that no player is updating has positive probability. This is to say, $M_{\mathbf{ss}} > 0$ for each state $\mathbf{s}$. Consequently, $S^{\tau-1} \subset S^\tau$ for each $\tau \geq 2$. The algorithm thus makes at most $|S|$ iterations. Since the set S is finite, the algorithm stops in finite time.

Clearly, if $S^0 \cup S^{\overline{\tau}} = S$ there are no recurrent sets aside from the absorbing states in S^0. To prove the opposite argument, assume that there are no recurrent sets aside from the absorbing states in S^0. At least one such absorbing state must be accessible from each state $\mathbf{s}$. Therefore, for some $\tau \leq \overline{\tau}$ and some absorbing state $\mathbf{s}'$, $M^\tau_{\mathbf{ss}'} > 0$, which implies that $\mathbf{s} \in S^\tau$, and thus $\mathbf{s} \in S^{\overline{\tau}}$. □

As it turns out every implementation of Algorithm 1 in this chapter yields the sets of absorbing states S^0 and of transient states $S^{\overline{\tau}}$, which satisfy $S^0 \cup S^{\overline{\tau}} = S$. Using proposition 6.4 we conclude that, across all the processes considered, the recurrent sets consist solely of absorbing states.

Throughout this chapter we assume that $q_i = q$ and $r_i = r$ for all i. For reference we include the payoff matrix of the prisoner's dilemma below.

		Player j C	D
Player i	C	c, c	e, f
	D	f, e	d, d

Without the loss of generality we may assume that all payoffs in the prisoner's dilemma game are positive, $f > c > d > e > 0$. We consider two outside option values, $o \in \{0, 40\}$. Using Proposition 6.2 it is straightforward to verify that the state $\mathbf{s}^{CE}$, in which all links are established, all players cooperate, and threaten to exclude any defective player, is an absorbing state of the basic process whenever

$$rc \geq f + (r-1)o, \tag{6.6}$$

i.e. whenever the payoff from r periods of mutual cooperation exceeds the payoff from one period of free-riding followed by $r-1$ periods of isolation. The complete characterization of absorbing states is, however, less straightforward. Below we analytically characterize absorbing states of the basic process with $o = 0$. In general, however, we use Algorithm 1 to characterize absorbing states of other processes considered in this chapter.

Let the value of the outside option be $o = 0$. For each player i the choice profile s_i^{DA} constitutes the unique dominant strategy in the basic game. Similarly, with any scope of forward looking r the state $\mathbf{s}^{DA}$, in which all links are established and all players defect, is an absorbing state of the basic process. For $r = 1$ absorbing states consist only of defective players. Proposition 6.5 below states that there might, however, exist absorbing states with cooperative players, when their scope of forward looking is sufficiently long. Note that in any state with type distribution $(p, n-p, 0, 0)$, such that $0 \leq p \leq n$, all n players are cooperative and $n-p$ are exclusive.

Proposition 6.5. *Consider the basic process with $o = 0$. State $\mathbf{s}^{DA}$ is absorbing for all r. State $\mathbf{s}^{DE}$ is absorbing for $r = 1$. States $\mathbf{s}$ with type distribution $z(\mathbf{s}) = (n-1, 0, 0, 1)$ are absorbing if and only if*

$$c > f - \frac{e}{n-2} + \frac{d}{r(n-2)} \,. \tag{6.7}$$

States $\mathbf{s}$ with type distribution $z(\mathbf{s}) = (p, n-p, 0, 0)$ are absorbing if and only if

$$c > \left(1 - \frac{(n-p-1)(r-1)}{(n-1)r}\right) f \,. \tag{6.8}$$

In particular, state $\mathbf{s}^{DE}$ is absorbing if and only if r satisfies (6.6). There are no other absorbing states.

The proof is given in Appendix A. Several general implications may be drawn from Proposition 6.5. Condition (6.6) with $o = 0$ is necessary for (6.8) and thus for the existence of any fully cooperative absorbing state. Condition

$$c \geq f - \frac{e}{n-2} \tag{6.9}$$

is necessary for the existence of absorbing states with type distribution $(n-1, 0, 0, 1)$. There is always an r such that (6.6) is satisfied, but this is not the case with (6.9). To summarize, when their scope of foresight is short (e.g. $r = 1$, myopic best response) all players are defecting in an absorbing state. Only when players have a sufficiently long scope of foresight do there exist absorbing states in which all players cooperate.

6.3.2 Stability of Absorbing States

The dynamics of the process may crucially depend on the initial conditions. In general, if a Markov chain starts in an absorbing state it will never leave that state. On the other hand, if it starts in a non-absorbing state it will leave that state and, in the long run, reach a recurrent set (or an absorbing state). The probability that the Markov chain is in a state that does not belong to a recurrent set is thus asymptotically zero. In this sense only the states in recurrent sets can be selected by the process in the long run.

It turns out that our basic process may have several absorbing sates but no other recurrent sets. The basic process reaches, with probability one, one of its absorbing states in a finite time and remains in that state forever. To get an indication about which of the absorbing states are most likely to be reached we introduce the notion of their stability. In short, some absorbing states may be more persistent than others, when the process is subject to small perturbations. If the players occasionally violate the updating rule, e.g. due to mistakes or experimentation, the process is most likely to spend most periods in an absorbing state that is robust against these violations.

This notion of stability has been formalized by concepts of stochastic stability [189] and the long-run equilibrium [190]. Both concepts have been invoked as solution criteria in evolutionary models with best response dynamics [191, 192]. Here, we use stochastic stability as the selection mechanism among the absorbing states of the basic process. An introduction to the concept of stochastic stability and discussion on the relation with concepts of asymptotic stability and evolutionary stability may be found in [193]. Essentially, stochastic stability requires robustness against persistent or regular mistakes, while evolutionary stability requires only robustness against isolated mutations. Young [194] argues in favor of its implementation in the studies of social phenomena and provides a series of applications of the concept.

Consider again the Markov chain defined by the best response dynamics introduced in Sec. 6.2. Assume also that the players occasionally make mistakes in the choice of their strategy. Let $\varepsilon \in (0,1)$ be the probability that an updating player chooses her new strategy at random. The player otherwise follows the best response rule (6.2). This modified updating rule generates a perturbed Markov process with state space S and transition matrix M^{ε}. Because of random error there is a positive probability $M^{\varepsilon}_{\mathbf{ss}'} > 0$ of a transition between any pair of states $\mathbf{s}, \mathbf{s}' \in S$, which implies that all entries in the transition matrix M^{ε} are positive. Proposition 2.1 describes a number of properties of a process with such transition matrix. In particular, the perturbed process is ergodic for each $\varepsilon > 0$, and has a unique stationary distribution $\mu^{\varepsilon}(\mathbf{s})$. This distribution fully characterizes the asymptotic properties of the process: independently of the initial state $\mathbf{s}^0$, the asymptotic proportion of time that the state *spends* in state $\mathbf{s}$ is given by $\mu^{\varepsilon}(\mathbf{s})$, and the probability that the process *is* in state $\mathbf{s}$ at time t approaches $\mu^{\varepsilon}(\mathbf{s})$ as t grows large.

Definition 6.1. *A state* $\mathbf{s}$ *is* stochastically stable *if* $\lim_{\varepsilon \to 0} \mu^{\varepsilon}(\mathbf{s}) > 0$.

The following Theorem is due to Young [189].

Theorem 6.1. *For each* $\mathbf{s} \in S$, $\mu^0(\mathbf{s}) = \lim_{\varepsilon \to 0} \mu^{\varepsilon}(\mathbf{s})$ *exists and the limiting distribution* μ^0 *is a stationary distribution of the non-perturbed process.*

It follows, in particular, that the basic process has at least one stochastically stable state. Intuitively, these are the states that are most likely to be observed over the long run when random mistakes are rare. The distribution μ^0 is one of solutions of the system (2.2). The stochastically stable sets therefore form a union of recurrent sets. However, it may not put positive probability on all recurrent sets. For the basic process this means that no other states than some of the absorbing states are stochastically stable.

6.4 Simulations of the Basic Process

To study dynamics of the process we employ two classes of computer simulations. Simulations of the first class are an implementation of Algorithm 1 and designed to detect the *direction of transitions* between the states of the Markov chain. Simulations of the second class are designed to study the *asymptotic properties* of the process, such as the stabilities of different absorbing states. The first class considers the non-perturbed basic process, while the second class considers its perturbed form. We consider two basic games. Game Γ^l coincides with the experimental game Γ^I in Chap. 5. Game Γ^h differs from that game only in the value of the outside option $o = 40$. The corresponding prisoner's dilemma game (5.1) and the updating parameters considered in the simulations are shown in Table 6.2. We provide the details of each class of simulations, and demonstrate their application, for the basic process with *low outside option* $o = 0$. We proceed by using the simulations to detect the dynamics and asymptotic properties of the basic process with *high outside option* $o = 40$.

Table 6.2. The parameters considered in the simulations of the basic dynamic process: (**a**) the network dilemma games; (**b**) the updating rule

$c = 50$	$e = 10$
$f = 70$	$d = 30$
outside option $o \in \{0, 40\}$	

(**a**)

probability of updating $q = 0.2$
scope of forward looking $r \in \{1, 2, 4\}$
number of players $n = 6$

(**b**)

6.4.1 Low Outside Option, $o = 0$

Consider $o = 0$. The parameters given in Table 6.2 do not comply with (6.9), but do comply with (6.6) for any $r \geq 2$. That is, there is no $r \in N$ such that states with type distribution $(n-1, 0, 0, 1)$ are absorbing, while fully cooperative absorbing states, for instance $\mathbf{s}^{CE}$, do exist for $r \geq 2$. According to Proposition 6.5 the absorbing states of the basic process, under these parameters, are

$r = 1 : \mathbf{s}^{DA}$ and $\mathbf{s}^{DE}$;
$r \in \{2, 3\} : \mathbf{s}^{DA}, \mathbf{s}^{CE}$ and all $\mathbf{s}$ such that $z(\mathbf{s}) \in \{(1,5,0,0), (2,4,0,0)\}$;
$r \geq 4 : \mathbf{s}^{DA}, \mathbf{s}^{CE}$ and all $\mathbf{s}$ such that
$z(\mathbf{s}) \in \{(1,5,0,0), (2,4,0,0), (3,3,0,0)\}$.

Below we discuss the dynamics of the basic process with scope of forward looking of 1, 2, or 4. These values of r were chosen because they are the smallest values representing all different sets of absorbing states.

6.4.2 Detecting Dynamics of the Process

The direct application of Algorithm 1 requires calculation of the transition matrix M of the non-perturbed process. In the worst case it will have $|S|$ iterations. With $n = 6$ and with 4 possible types the state space S consists of $4^6 = 4096$ states and the transition matrix M of $4096^2 > 16.000.000$ entries. Algorithm 1 would in the worst case require 4096 operations on the matrix M.

A simplification of Algorithm 1 is possible, however. According to Proposition 6.2 it is sufficient to classify one of the states with type distribution $z \in Z$ as absorbing (transient) to prove that every state with type distribution z is absorbing (transient). Further, to prove that a state $\mathbf{s} \in S$ is transient it is sufficient to simulate the updating process, initiated with $\mathbf{s}[1] = \mathbf{s}$, until an absorbing state or another transient state is reached. If, on the other hand, the process makes no transitions for many periods from the state $\mathbf{s}$, it is likely to be absorbing.

We employ these observations and use computer simulations for detection of the absorbing states and of the direction of transitions from the transient states. For each possible initial distribution $z \in Z$ we choose a state $\mathbf{s}_z \in S_z$ and simulate 100 iterations of the non-perturbed basic process initiated with $\mathbf{s}[1] = \mathbf{s}_z$. Additionally, for each $\mathbf{s}_z$ we repeat the simulation 100 times.

If no transition from the initial state is observed during the 100 periods then we classify it as absorbing. The probability of misclassification is below 10^{-4}. To see this, note that in a non-absorbing state at least one player would change her type when updating. For the process to remain in the non-absorbing state, such player should never update. The probability that a player never updates during 100 periods is $(1-q)^{100} = 0.9^{100} < 10^{-4}$. Further, by repeating each such simulation 100 times, we assure observations of many transitions from every non-absorbing state. The probability that we observe less than 100 such transitions is $(1-10^{-4})^{100} < 0.01$.

In the course of simulations we progressively classify the states as follows. We begin by classification of absorbing states. We classify any remaining state as transient whenever the process initiated in that state reaches an absorbing state. We proceed by iteration and classify any state as transient whenever the process initiated in that state reaches another transient state. If, after a number of such iterations, all states are classified as either absorbing or transient we conclude that there are no recurrent sets other than absorbing states.

By following the transitions of the processes across all initial states $\mathbf{s}_z$ we can also get an impression of the dynamics of cooperation and exclusion in the basic process. To turn this impression into a two-dimensional graphical presentation we must aggregate the information about the observed transitions. For each state $\mathbf{s} \in S$ let $y_1(\mathbf{s}) = n^{CA}(\mathbf{s}) + n^{CE}(\mathbf{s})$ be the number of cooperative players and let $y_2(\mathbf{s}) = n^{CE}(\mathbf{s}) + n^{DE}(\mathbf{s})$ be the number of exclusive players in the state $\mathbf{s}$. Each realization of the basic process $\mathbf{s}[t]$ implies a realization of the truncated process

$$y[t] = (y_1(\mathbf{s}[t]), y_2(\mathbf{s}[t]))$$

representing the corresponding evolution of cooperation and exclusion. The process $y[t]$ is the projection of the basic process onto the two dimensional state space $Y = \{0, 1, ..., 6\}^2$.

For each $y \in Y$ we characterize how cooperation and exclusion change in one period, starting from a state with y_1 cooperative and y_2 exclusive players. For this we estimate, for each y, the average value of $y[t+1]-y[t]$ conditional on $y[t]=y$. Let this value be given for each $y \in Y$ by the mapping $\chi : Y \to \mathbb{R}^2$. For each $y \in Y$ the value of $\chi(y)$ gives the average observed *direction of transitions* from y. Specifically, the value of $\chi_1(y)$ gives the average increase in cooperation, and the value of $\chi_2(y)$ the average increase in exclusion, observed between a period in which there are y_1 cooperative and y_2 exclusive players and the next period. In other words, the states with y_1 cooperative and y_2 exclusive players induce an average increase in cooperation of $\chi_1(y)$ and an average increase in exclusion of $\chi_2(y)$.

For completeness we include the formal definition of the mapping χ. Let Ψ be the set of observed realizations of the basic process. For each $y \in Y$ and realization $\mathbf{s}[t]_{t=1}^{100} \in \Psi$, let

$$\mu(y \mid \mathbf{s}[t]_{t=1}^{100}) = |\{t \leq 100 \mid (y_1(\mathbf{s}[t]), y_2(\mathbf{s}[t])) = y\}|$$

be the number of periods t during which the realization $\mathbf{s}[t]_{t=1}^{100}$ was in a state with y_1 cooperative and y_2 exclusive players. The mapping χ is then defined for each $y \in Y$ by

$$\chi(y) = \sum_{\substack{\mathbf{s}[t]_{t=1}^{100} \in \Psi \\ y(\mathbf{s}[t])=y}} (y(\mathbf{s}[t+1]) - y(\mathbf{s}[t])) \Bigg/ \sum_{\mathbf{s}[t]_{t=1}^{100} \in \Psi} \mu(y \mid \mathbf{s}[t]_{t=1}^{100}) \ .$$

The mapping χ does not accurately describe the basic process. The main shortcoming of using a projection of the process is that it might aggregate over states with very different dynamic properties. It does, however, offer a fair idea of the dynamics within the process.

Results of this class of simulation are presented in Fig. 6.1 for $r = 1, 2, 4$. The dark spots represent absorbing states of the process. The arrows represent the mapping χ. For each $y \in Y$, $\chi_1(y)$ determines the vertical dimension and $\chi_2(y)$ the horizontal dimension of the corresponding arrow. In this respect the length and the direction of the vectors represent the "speed" and "direction" of the dynamics. If for some $y \in Y$ all corresponding states are absorbing, only a dark spot is shown but no arrow.

For all values of $r = 1, 2, 4$ the sets of absorbing states yielded by the simulations coincide with those given by Proposition 6.5. All the remaining states are transient which implies that there are no recurrent sets, other than absorbing states. An inspection of Fig. 6.1 suggests that the dynamics in the basic process overwhelmingly favors state $\mathbf{s}^{DA}$ over alternative absorbing states. The figure also gives an indication about the dynamics of the perturbed process that we discuss in detail in the

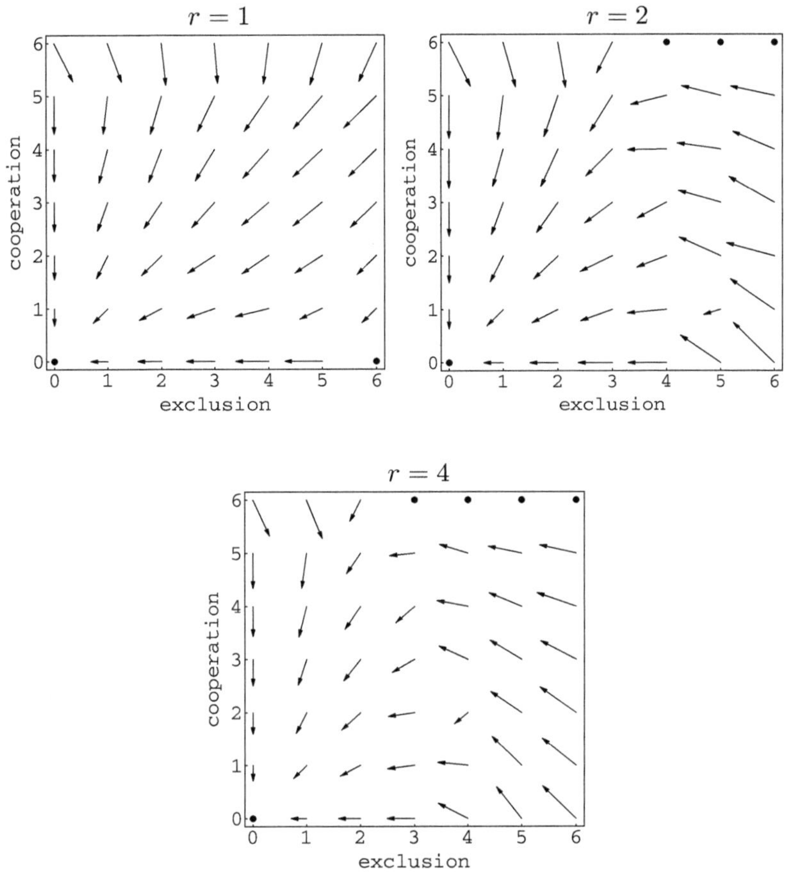

Fig. 6.1. Basic process with o=0. In this and subsequent figures the dynamics of various *non-perturbed* processes are shown for $r = 1, 2, 4$. The interpretation of the vector plots is given in the text. Briefly, given a state with y_1 cooperative and y_2 exclusive players the arrow that originates at the corresponding location shows the relative likelihood and the average direction of the transition from this state. The dark spots illustrate the absorbing states

following section. For $r = 2$ and $r = 4$ a single mistake in the choice of best response may lead the process away from absorbing states with full cooperation and into the state $\mathbf{s}^{DA}$ with full defection. In contrast, once in $\mathbf{s}^{DA}$, the process returns to $\mathbf{s}^{DA}$ even after several simultaneous mistakes. In this sense the state $\mathbf{s}^{DA}$ is the only stable among the absorbing states. The direction of the dynamics suggests that, starting from a random initial state, $\mathbf{s}^{DA}$ is the most likely absorbing state to be selected by the process over the long run, for all considered values of r.

6.4.3 Detecting the Asymptotic Behavior of the Process

Using the first class of simulations we characterized the absorbing states of the basic process, and the dynamics between the remaining states. The process considered was non-perturbed and assumed that players always follow the updating rule (6.2) precisely. This assumption allows us to obtain a useful benchmark for the asymptotic properties of the process, such as its absorbing states. To select among the absorbing states, and to study the basic process in the presence of perturbations, we consider the second class of simulations. We still assume that players follow their best response most of the time, but allow occasional deviations from the updating rule (6.2). That is, when updating a player chooses a random strategy with small probability ε, and follows the rule (6.2) otherwise.

The asymptotic properties of the slightly perturbed process may be of interest for various reasons. Firstly, human players are often prone to errors in their decision making. Deviations may be unintentional, e.g. due to imperfect information or limited computing abilities, or intentional, such as when players choose to play a non-optimal strategy hoping to induce a transition from one equilibrium to another. In the setting of the repeated basic game, human players may, for example, realize that the absorbing state $\mathbf{s}^{DA}$ is payoff dominated by the absorbing state $\mathbf{s}^{CE}$ and choose to cooperate and exclude defectors with intention to induce a transition between the two absorbing states. The perturbed process may give us an impression on how successful players can be in inducing such transitions.

Secondly, even though the recurrent sets are the only candidates for the long run selection by the non-perturbed process, the perturbed process may spend only a small fraction of time in those sets. For instance, we see below that the perturbed basic process with $o = 0$ spends most of the periods in the absorbing state $\mathbf{s}^{DA}$. It spends most of the remaining periods in the states with type distributions $\{(1, 0, 5, 0), (0, 1, 5, 0), (0, 0, 5, 1)\}$, which arise by a single deviation from the absorbing state $\mathbf{s}^{DA}$ and have a high number of defective players. We denote a set of these states by $\{\mathbf{s}^{DA} - 1\}$.

However, in some instances even a small perturbation of the process results in dynamics that only rarely stabilizes in an absorbing state. In Sec. 6.4.4, for example, we show that the perturbed basic process with $o = 40$ and $r = 2$ spends a significant number of periods in states with many cooperative players, even though all absorbing states are fully defective.

Finally, because the considered a small error probability, the stationary distribution $\mu^{\varepsilon}(s)$ gives us an indication about the limit distribution $\mu^0(s) = \lim_{\varepsilon \to 0} \mu^{\varepsilon}(s)$ and the stochastically stable states. In Sec. 6.3.2 we argue that only the absorbing states of the non-perturbed process may be stochastically stable. The comparison of the values of $\mu^{\varepsilon}(s)$ between the absorbing states of the non-perturbed process thus allows us to speculate about their stability.

In Sec. 6.3.2 we argue that the perturbed basic process is ergodic, with a unique stationary distribution $\mu^{\varepsilon}(s)$. Analytical computation of $\mu^{\varepsilon}(s)$ requires computation of matrix M^{ε} and finding the solution of system (2.2), given this matrix. If the process is observed for a large number of periods τ, however, the observed frequency distribution $\mu_{\varepsilon}^{\tau}(\mathbf{s} \mid \mathbf{s}^0)$ gives with high probability a close approximation of $\mu^{\varepsilon}(s)$, independently of the initial state $\mathbf{s}^0$. Instead of computing $\mu^{\varepsilon}(s)$ analytically we may therefore follow the perturbed process, initiated in an arbitrary initial state, and, after some large number of periods τ, consider $\mu_{\varepsilon}^{\tau}(\mathbf{s} \mid \mathbf{s}^0)$ as a good approximation of $\mu^{\varepsilon}(s)$. The calculation of the convergence rate of the frequency distribution toward the stationary distribution requires equal computing capacity as the analytical calculation of $\mu^{\varepsilon}(s)$.[6] Here, we rely rather on an intuitive criterion to decide how long to follow the process.

We simulate the process sufficiently long to observe a sufficient number of transitions between the most distant states. The states which, between them, require the maximal number of transitions are the corner states $\mathbf{s}^{CA}$, $\mathbf{s}^{CE}$, $\mathbf{s}^{DA}$ and $\mathbf{s}^{DE}$. The criterion we implement is to continue the simulations until at least 5 transitions are detected into each of the corner states, from another corner state. The process is initiated in the state $\mathbf{s}^0 = \mathbf{s}^{CA}$. We consider $\varepsilon = 0.1$.[7] The values of $\mu^{\tau}(\mathbf{s} \mid \mathbf{s}^0)$ for the most frequent states of the non-perturbed basic pro-

[6] Convergence in probability may be evaluated as follows (see [40] for details). For any two distributions ϱ and σ over the states of the process consider $\|\varrho - \sigma\| = \max_{x \in X} |\varrho_x - \sigma_x|$, the maximal difference in probabilities between two distributions. For an ergodic Markov chain on a finite state space, with transition matrix M and stationary distribution μ, the convergence of ϱM^t awards μ is at an exponential rate for any initial distribution ϱ, that is, $\|\varrho M^t - \mu\| \leq c(\theta^t)$ for some $0 < \theta < 1$ and $c > 0$. To find θ one needs to calculate the largest non-unit eigenvalue of M, calculation of which is comparable to solving the system $\mu M = \mu$. This property can then be used to compare rates of convergence between local and global interaction processes [173].

[7] Smaller error probability may give a better indication about stochastic stability of absorbing states but also substantially increases the duration of simulations.

Table 6.3. Basic process with o=0. In this and subsequent tables the left panel shows relative frequencies of absorbing and most frequent other states, while the right panel shows relative frequencies of different types, all obtained from long-run simulations of various *perturbed* processes, for $r = 1, 2, 4$. In these tables "k coop." represents all states with k cooperative players

states r:	1	2	4
6 coop.	.01	.01	.01
$\mathbf{s}^{DA}$	.65	.65	.64
$\{\mathbf{s}^{DA} - 1\}$	.29	.28	.29

type r:	1	2	4
CA	.02	.02	.02
CE	.02	.02	.02
DA	.93	.93	.93
DE	.03	.03	.03

cess, as well as the frequencies of different types, are reported in Table 6.3. For any value of r the process spends around 65 percent of periods in state $\mathbf{s}^{DA}$ and around 29 percent of periods in states $\{\mathbf{s}^{DA} - 1\}$ that can be reached from state $\mathbf{s}^{DA}$ by a deviation of a single player. We conclude that for all considered values of r the state $\mathbf{s}^{DA}$ appears to be unique stochastically stable state. Consequently, DA is by far the most frequently observed type.

6.4.4 High Outside Option, $o = 40$

To study the dynamics and the long run behavior of the basic process with $o = 40$ we employ the same classes of simulations as in the case with $o = 0$. We do not provide an analytical characterization of the absorbing states for $o = 40$. Rather, we rely on the simulations of the first class to accurately classify the sets of absorbing and transient states. We have argued above that an inaccurate classification of the states is very unlikely.

Again, we consider parameters given in Table 6.2, with $o = 40$. The absorbing states of the basic process, as classified by the first class of simulations under these parameters (see also Fig. 6.2), are

$$\begin{aligned} r = 1: &\ \mathbf{s}^{DE}, \text{ and } \mathbf{s} \text{ such that } z(\mathbf{s}) \in \{(0,0,1,5), (0,1,0,5), (0,1,1,4)\}, \\ r = 2: &\ \mathbf{s}^{DE}, \text{ and } \mathbf{s} \text{ such that } z(\mathbf{s}) = (0,0,1,5), \\ r = 4: &\ \mathbf{s}^{CE}, \mathbf{s}^{DE}, \text{ and } \mathbf{s} \text{ such that } z(\mathbf{s}) = (0,0,1,5). \end{aligned}$$

All the remaining states were classified as transient, which implies that there are no recurrent sets, other than absorbing states. In the state $\mathbf{s}^{DE}$ all players defect and exclude other defective players, which results in an empty network with isolated defective players. For convenience we denote by $\{s^{DE} + 1\}$ the set of states with type distributions in $\{(0,0,1,5), (0,1,0,5), (0,1,1,4)\}$.

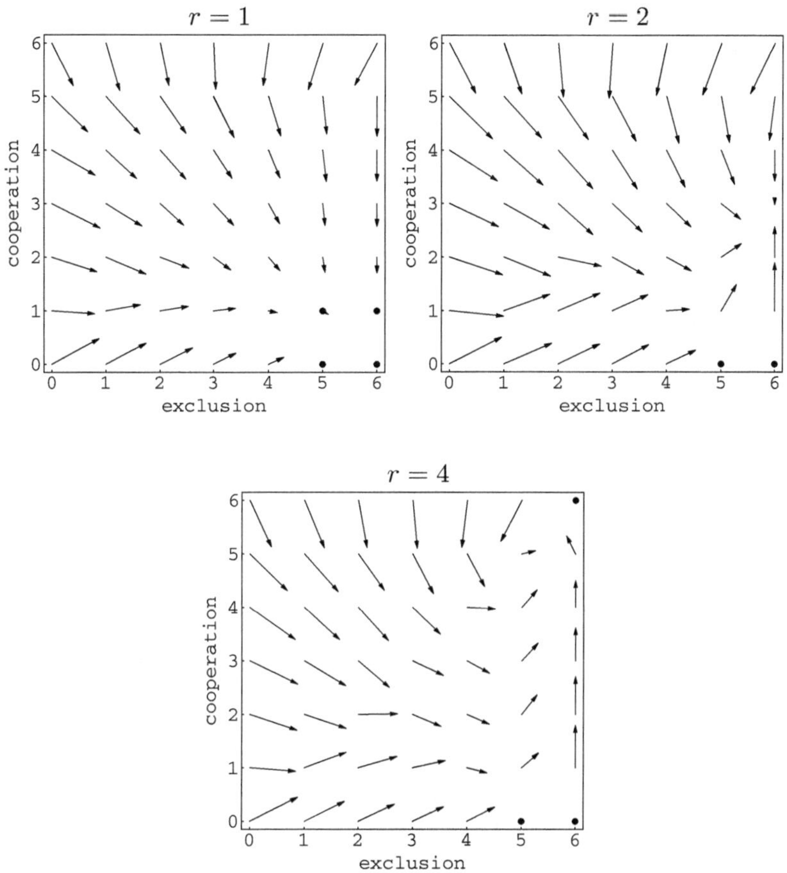

Fig. 6.2. Basic process with o=40

Results of this class of simulation are presented in Fig. 6.2 for $r = 1, 2, 4$. The direction of the dynamics for $r = 1$ suggests that, starting from a random initial state, the process quickly reaches an absorbing state. For $r = 2$ only fully defective absorbing states exist but the dynamics suggests that the process may need a large number of periods to reach one of them (state $(0, 3, 0, 3)$ is not absorbing). More interesting, the illustration suggests that small perturbations may lead the process to spend a substantial proportion of periods in states with 2, 3 or 4 cooperative players. For $r = 4$ the figure does not obviously suggest which of the absorbing states is most likely to be selected by the process. Judging by the dynamics close to the absorbing states, state $\mathbf{s}^{CE}$ seems to be most attractive.

Table 6.4. Basic process with o=40. Notation "other: k excl." represents those states with k exclusive players not mentioned in other categories

states r:	1	2	4
$\mathbf{s}^{CE}$	.01	.01	.27
$\mathbf{s}^{DE}$	.09	.10	.03
$\{\mathbf{s}^{DE}+1\}$	.69	.20	.07
other: 6 excl.	.02	.45	.21

type r:	1	2	4
CA	.02	.02	.05
CE	.11	.34	.64
DA	.12	.06	.07
DE	.75	.58	.25

The observations above can be verified using the second class of simulations. Again, the perturbed process with $\varepsilon = 0.1$ is initiated in the state $\mathbf{s}^0 = \mathbf{s}^{CA}$. For convenience we refer to the absorbing states of the non-perturbed process simply as absorbing states. The values of $\mu^{\tau}(\mathbf{s} \mid \mathbf{s}^0)$ for absorbing states and most frequent transient states, as well as the frequencies of different types, are reported in Table 6.4. The statistics about the asymptotic behavior of the process confirm our conjectures inferred from Fig. 6.2. For all considered values of r exclusion is very high. For $r = 1$ the perturbed process spends most of the time in the absorbing states implying low frequency of cooperation and low density of links. For $r = 2$, however, the perturbed process spends almost half of the time in the non-absorbing states with all players exclusive and some of them cooperative. For $r = 4$ the absorbing state $\mathbf{s}^{CE}$ in which all players cooperate and threaten to exclude defectors, appears the unique stochastically stable state.

It is interesting to conclude that with $o = 0$ the process remains in state $\mathbf{s}^{DA}$ with full defection for most of the time even when there is a small probability that players experiment with sub-optimal strategies. This contrasts with the behavior of the process with $o = 40$ where some cooperation appears as long as the players have a non-trivial scope of forward looking $r \geq 2$. In this case a single cooperator may disturb the defective absorbing state $\mathbf{s}^{DE}$ by inducing even more players to cooperate. When their scope of forward looking is $r = 4$ players remain in fully cooperative and exclusive state $\mathbf{s}^{CE}$ for much of the time.

6.5 Social Preferences

In the previous section players were assumed to consider only their own payoff when choosing their strategy. In this sense we may say they were *selfish* for not considering the consequences of their choices for the payoffs of the remaining players in the group. When shortsighted such players do not fare well in terms of cooperation. With scope of forward looking $r = 1$, and players choosing their strategy by maximizing their own immediate payoff, cooperation could not be sustained in the long run.

This is not surprising in view of the theoretical results that cooperation in social dilemma games is unlikely if selfish players behave according to the myopic best response. In Chap. 4, for example, we show that there should be no cooperation in the finitely repeated basic game if it is (common knowledge that it is) played by fully rational selfish players. Under these assumptions the same holds also for general public goods games with or without punishment [95].

These predictions, however, are in stark contrast with experimental evidence. In Chap. 5 we discuss an experimental study of the finitely repeated basic game. The observed average cooperation rates in the experimental game with partner choice and no linking constraints were about 70%. Similarly, positive contributions are commonly observed in experiments with public goods games [195]. If a costly punishment possibility is included with the public goods game the contribution rates even increase, may be sustained until the last periods, and may reach 100% [96].

One explanation provided by several authors is that a non-negligible proportion of people are not really selfish. The utility of people may depend on their own payoff as well as on the payoff of other players in their reference group. We say that people with such utilities are *other-regarding* and exhibit social preferences rather than selfish ones. In this section we look at two models of social preferences: the model of altruism by Levine [132] and the model of inequity aversion by Fehr and Schmidt [95]. Each has been shown to explain positive contributions in public goods games with punishment, even when assuming myopic best response behavior. In Chap. 5 we study equilibria of our basic game if played by other-regarding, but rational players. In this section we look at the dynamics of cooperation and exclusion in the repeated basic game, played by other-regarding players with limited forward looking. For each set of parameters considered we characterize dynamic and asymptotic properties of the corresponding processes using the simulations introduced in Sec. 6.4. For convenience we repeat the essential definitions from Chap. 5.

In Tables 6.5–6.8 below we use the following notation:

"6 (5) coop." denotes the set of all states with 6 (or 5) cooperative players;

"other: 6 (5) excl." denotes the set of states with 6 (or 5) exclusive players (except those already included in other categories);

"$\{\mathbf{s}^{DA}-1\}$" denotes the set of states with type distribution $(1,0,5,0)$, $(0,1,5,0)$ or $(0,0,5,1)$;[8]
"$\{\mathbf{s}^{DE}+1\}$" denotes the set of states with type distribution $(0,1,0,5)$, $(0,0,1,5)$, or $(0,1,1,4)$.[9]

6.5.1 Process with Altruism

To explain, i.a., positive contributions in public goods games Levine [132] proposes a model of altruism, which assumes that players gain additional utility from payoffs to those other players they consider nice, and lose utility from payoffs to those other players they consider aggressive. Each player is associated with a *coefficient of altruism* $\gamma_i \in (-1,1)$ which determines her other-regarding inclination. If payoffs of players are given by the vector $(x_1,...,x_n)$, player i receives the (adjusted) utility

$$u_i^A = x_i + \sum_{j\neq i} \frac{\gamma_i + \lambda\gamma_j}{1+\lambda} x_j \ ,$$

where the coefficient $\lambda \in [0,1]$ reflects the fact that players have a higher regard for altruistic opponents than for the spiteful ones.

In the basic process all players were assumed to consider only their own payoff, which is the special case of the model with altruism with trivial coefficients $\gamma_i = 0$ for all i. Levine, however, estimates that data from ultimatum game experiments [104] are better explained if $\lambda = 0.45$. Levine further shows that the data from the public goods game experiments [134] are roughly consistent with the following distribution of coefficients of altruism:

$$\gamma_i = \begin{cases} -0.9\ ;\ pr = 0.20 \\ -0.22\ ;\ pr = 0.52 \\ 0.133\ ;\ pr = 0.04 \\ 0.243\ ;\ pr = 0.17 \\ 0.453\ ;\ pr = 0.07 \end{cases} . \qquad (6.10)$$

In the following we study the adaptation of the basic process by considering players, whose coefficients of altruism are non-trivial. We consider a group of n other-regarding players with a uniform scope of forward looking r repeatedly playing the basic game. At each time t

8 "$\{\mathbf{s}^{DA}-1\}$" represents states that can be reached from state $\mathbf{s}^{DA}=(0,0,6,0)$ by a deviation of one player.

9 "$\{\mathbf{s}^{DE}+1\}$" represents absorbing states of the processes with high outside option and $r=1$.

each player i plays one of the strategies $\{s_i^{CA}, s_i^{CE}, s_i^{DA}, s_i^{DE}\}$. As in the basic process, at each time $t \geq 2$, before making her move, player i with some fixed probability $q_i \in (0,1)$ updates her own strategy, by observing the vector $\mathbf{s}[t-1]$ of strategies at time $t-1$. A player updates by choosing a strategy which maximizes her utility across the next r_i periods, under the assumption that the strategies of the other players do not change in those periods.

Formally, let $\mathbf{s} = (s_1, ..., s_n) \in S$ be the vector of strategies at time $t-1$, and let $\mathbf{s}' = (s_1, ..., s_{i-1}, s_i', s_{i+1}, ..., s_n)$ be the vector of strategies at time t. Player i changes her strategy from s_i to s_i' at t and assumes no other updates in the group during $[t-1, t+r_i]$. Let $x_i(s_i' \mid \mathbf{s})$, defined as in (6.1), be the payoff of player i across the r_i periods following her update, and let

$$\begin{aligned} x_{ij}(s_i' \mid \mathbf{s}) &= \pi_j(a_j(s_j), p_j(s_j \mid a(\mathbf{s})), a_{-j}(\mathbf{s}'), p_{-j}(\mathbf{s}' \mid a(\mathbf{s}))) \\ &\quad + (r_i - 1)\pi_j(a_j(s_j), p_j(s_j \mid a(\mathbf{s}')), a_{-j}(\mathbf{s}'), p_{-j}(\mathbf{s}' \mid a(\mathbf{s}'))), \quad (6.11) \end{aligned}$$

be the payoff to player j across the same periods. The utility of player i, for changing her strategy s_i to strategy s_i' is then given by

$$u_i^A(s_i' \mid \mathbf{s}) = x_i(s_i' \mid \mathbf{s}) + \sum_{j \neq i} \frac{\gamma_i + \lambda \gamma_j}{1 + \lambda} x_{ij}(s_i' \mid \mathbf{s}) \ . \qquad (6.12)$$

If player i is updating at the start of period t she chooses a strategy $s_i[t]$ such that

$$s_i[t] \in \arg\max_{s_i' \in S_i} u_i^A(s_i' \mid \mathbf{s}[t-1]) \ .$$

If this payoff is maximized by the current strategy of the player, the player keeps her current strategy. If not, she randomly selects one of the optimal strategies.

The updating rule outlined above generates a Markov chain with the finite state space S and transition matrix M^A. We refer to this Markov chain the *process with altruism*. The basic process can be seen as a special case of the process with altruism when all coefficients of altruism are trivial. Using the same argument that proved Proposition 6.1 it can be shown that in the process with altruism a state $\mathbf{s}$ is absorbing if and only if

$$s_i \in \arg\max_{s_i' \in S_i} u_i^A(s_i' \mid \mathbf{s}) \ , \qquad (6.13)$$

for each player i. Propositions 6.3 and 6.4 hold for any Markov chain and thus for the process with altruism. The following proposition is a generalization of Proposition 6.2 for the process with altruism. Its proof is given in Appendix A.

Proposition 6.6. *Consider the process with altruism and let there be* γ *such that* $\gamma_i = \gamma$ *for each player* i*. Let* $z \in Z$ *be a type distribution. State* $\mathbf{s} \in S_z$ *is absorbing if and only if each of the states in* S_z *is absorbing. State* $\mathbf{s} \in S_z$ *is transient if and only if each of the states in* S_z *is transient.*

In Sec. 6.4.2 we argue that Proposition 6.2 permits us to use the first class of simulations, instead of Algorithm 1, to detect the absorbing and transient states of the basic process. With Proposition 6.6 we can argue in the same way that, whenever players are homogeneous with respect to their coefficients of altruism, we may implement the first class of simulations to detect absorbing and transient states of the process with altruism.

We study two cases with players having homogeneous coefficients of altruism, and one case with players whose coefficients of altruism are distributed according to distribution (6.10). The three processes are characterized by:

A1 $\gamma_i = 1$ for each player i,
A2 $\gamma_i = 0.453$ for each player i, or
A3 for each player i, γ_i is independently drawn from (6.10).

We assume $r_i = r$ and $q_i = q$ for each i. We consider all the stage game constellations given by Table 6.2 to achieve consistency with parameters considered in the basic process.

We want to study the process with altruism both with realistic coefficients of altruism, as in A3, and with those that give the best chance of cooperation. We conjecture that players with high coefficients of altruism are more likely to achieve cooperation than those with low coefficients of altruism. With this reason we consider the highest feasible coefficient of altruism in A1 and the highest coefficient of altruism allowed by (6.10) in A2.

Processes A1 and A2 consider homogeneous coefficients of altruism and may be studied via simulations of the first, non-perturbed, class or the second, perturbed, class. Process A3 considers heterogeneous coefficients of altruism and may be studied only via the simulations of the second class. Moreover, (6.10) induces a distribution over possible allocations of coefficients in the group. To study the asymptotic behavior of process A3 we repeat 50 runs of the following procedure. In each run we independently draw from (6.10) a coefficient of altruism for each

Table 6.5. Processes with altruism with o=0. Perturbed

A1

states	r: 1	2	4
6 coop.	.69	.70	.69
5 coop.	.25	.24	.24

type	r: 1	2	4
CA	.78	.79	.79
CE	.16	.15	.15
DA	.03	.03	.03
DE	.03	.03	.03

A2

states	r: 1	2	4
6 coop.	.01	.06	.06
$(5,0,0,1)$	.01	.13	.13
$\mathbf{s}^{DA}$	.65	.32	.32
$\{\mathbf{s}^{DA}-1\}$	.29	.17	.17

type	r: 1	2	4
CA	.02	.33	.33
CE	.02	.04	.04
DA	.93	.58	.58
DE	.03	.05	.05

A3

states	r: 1	2	4
6 coop.	.01	.01	.01
$\mathbf{s}^{DA}$	.65	.64	.63
$\{\mathbf{s}^{DA}-1\}$	.29	.29	.29

type	r: 1	2	4
CA	.02	.02	.03
CE	.02	.02	.02
DA	.93	.93	.92
DE	.03	.03	.03

player, and simulate 1.000.000 periods of the perturbed process using simulations of the second class. We assume that coefficients of altruism are common knowledge. To report the results of these simulations we aggregate the frequencies observed across all 50 runs.

Low Outside Option

Let $o = 0$. For each value of r the dynamics of A1 is shown in the left column and the dynamics of A2 in the right column of Fig. 6.3. In Table 6.5 we report the results of long run simulations of perturbed processes A1, A2 and A3. For each of the processes we note the frequencies of the most frequent states and the frequency distribution across different types.

By looking at Fig. 6.3 we may conclude that cooperation is likely only among sufficiently altruistic players. This seems certainly true when players are altruistic at the extreme, as in the process A1. However, players with high, but not extreme, coefficients of altruism also need to be sufficiently forward looking to achieve cooperation. As is

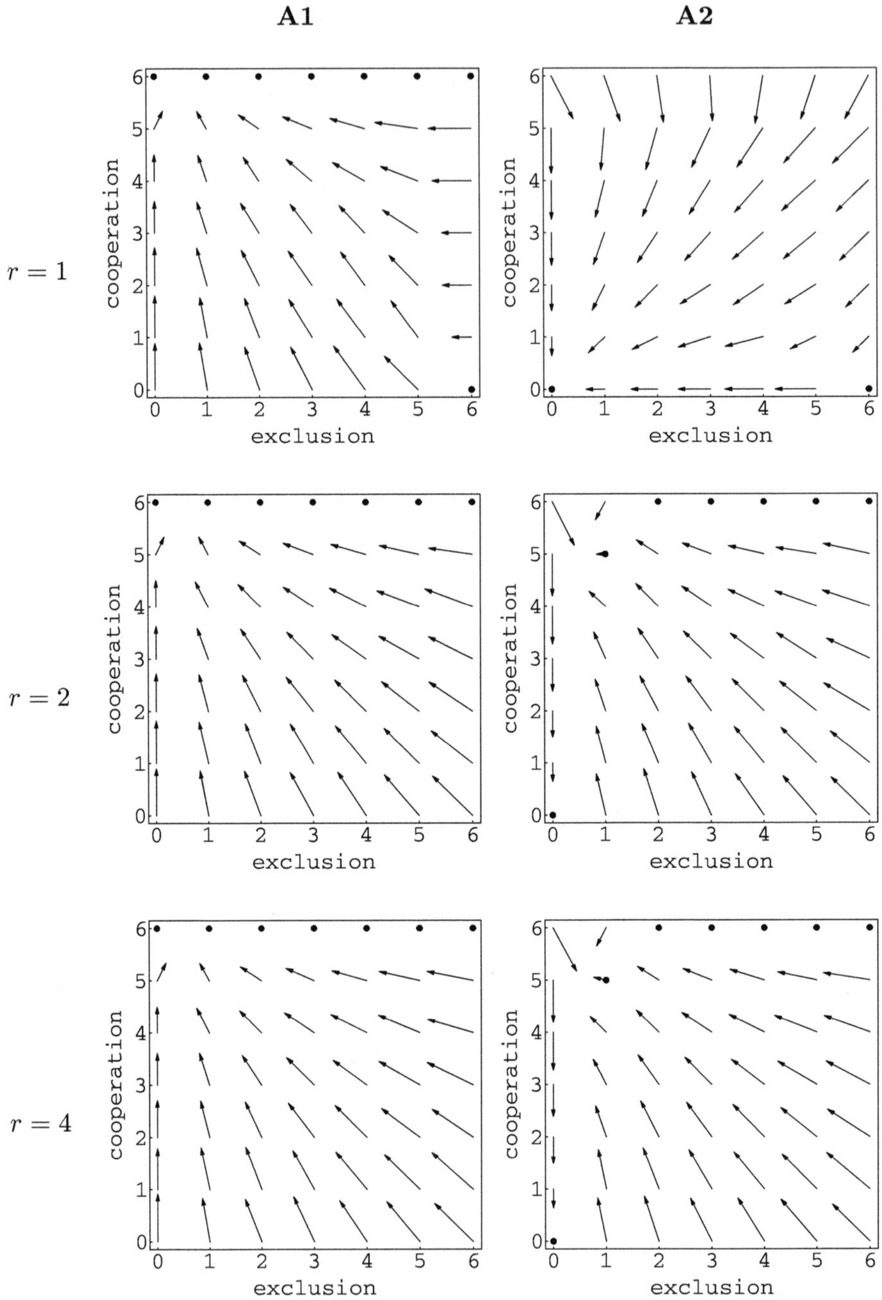

Fig. 6.3. Processes with altruism with o=0. Non-perturbed

evident from Fig. 6.3(b) players whose coefficients of altruism are maximal among those suggested by (6.10) fail to attain cooperation when $r = 1$, that is, if they behave according to the myopic best response.

These observations are confirmed by the second class of simulations, see Table 6.5. On the long run cooperation is sustained only in the process A1 with players that have extremely high coefficients of altruism, e.g. if $\gamma_i = 1$ for all players. Consider now the process A2 with players whose coefficients of altruism are the highest among those suggested by the theory. Cooperation can be achieved among such players with positive probability only if they are sufficiently forward looking, that is, if $r \geq 2$. If $r = 1$ then the state $\mathbf{s}^{DA}$, in which all players defect, seems to be the unique stochastically stable state and DA is the most frequently observed type. The results for the process A3 show that cooperation cannot be sustained if coefficients of altruism are distributed as suggested by the model of altruism. The process A3 spends around 65 percent of periods in state $\mathbf{s}^{DA}$ and around 29 percent of periods in states $\{\mathbf{s}^{DA} - 1\}$. The state $\mathbf{s}^{DA}$ seems to be the unique stochastically stable state and DA is the most frequently observed type for all values of r.

High Outside Option

Let $o = 40$. For each value of r the dynamics of A1 is shown in the left column and the dynamics of A2 in the right column of Fig. 6.4. In Table 6.6 we report the results of the long run simulations of processes A1, A2 and A3, respectively. For each of the processes we note the frequencies of the most frequent states and the frequency distribution across different types.

Again, by looking at Fig. 6.4 and at the frequencies given in Table 6.6, we may conclude that cooperation is likely only among sufficiently altruistic or sufficiently forward looking players. In the process A1 with players with extremely high coefficients of altruism cooperation always prevails. In the process A2 with players whose coefficients of altruism are the highest among those suggested by Levine's model complete cooperation can be sustained if $r \geq 2$ but not if $r = 1$. Finally, in the process A3 where coefficients of altruism are distributed as suggested by the model of altruism, cooperation increases with forward looking. If $r = 4$ then 52 percent of observed types are cooperative.

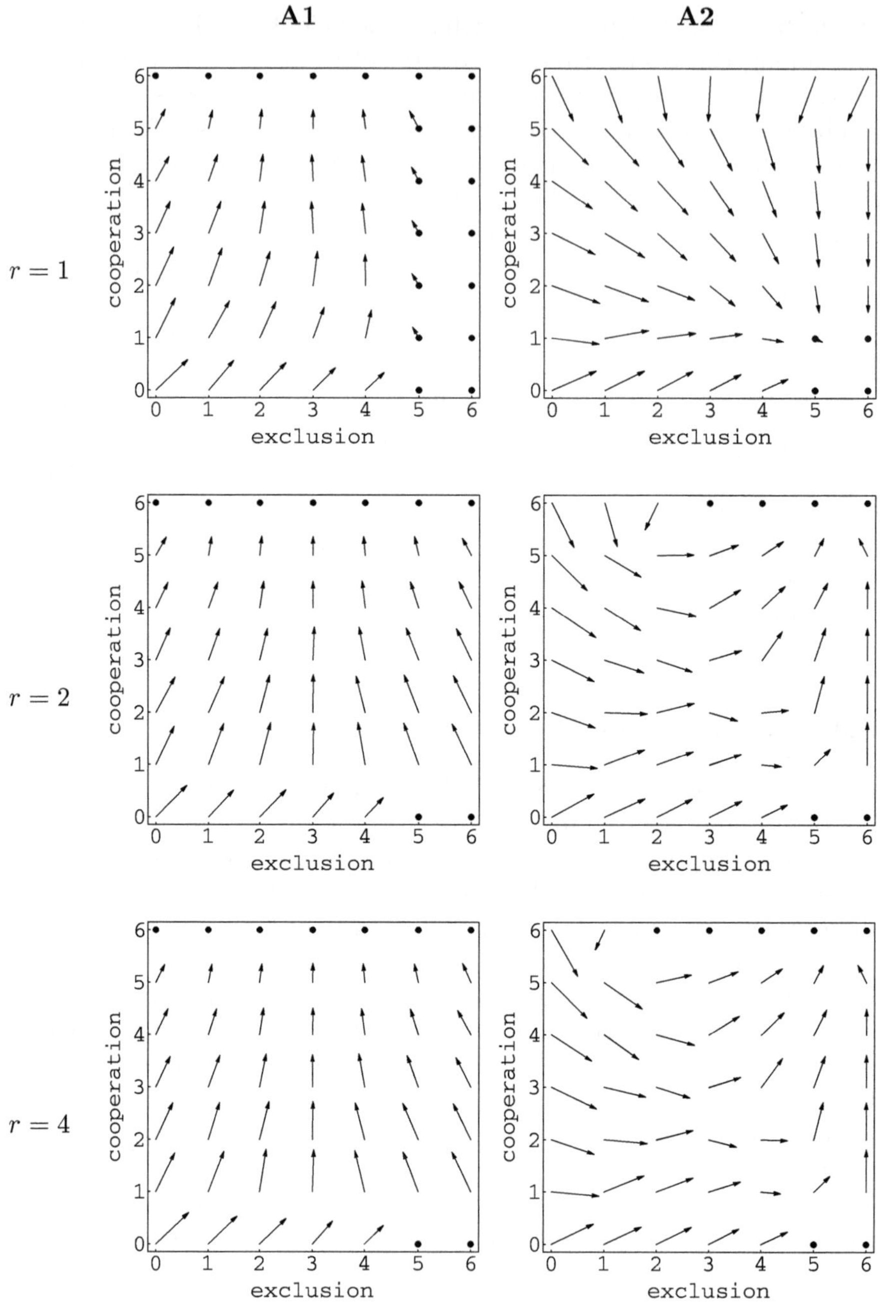

Fig. 6.4. Processes with altruism with o=40. Non-perturbed

Table 6.6. Processes with altruism with o=40. Perturbed

A1

states	$r:$ 1	2	4
6 coop.	.65	.68	.68
5 coop.	.26	.24	.24

type	$r:$ 1	2	4
CA	.46	.47	.47
CE	.46	.47	.47
DA	.04	.03	.03
DE	.04	.03	.03

A2

states	$r:$ 1	2	4
6 coop.	.01	.68	.69
$\mathbf{s}^{DE}, \{\mathbf{s}^{DE}+1\}$	.78	.01	.01
other: 6 excl.	.02	.08	.08

type	$r:$ 1	2	4
CA	.02	.15	.15
CE	.11	.79	.79
DA	.12	.03	.03
DE	.75	.03	.03

A3

states	$r:$ 1	2	4
$\mathbf{s}^{CE}$	.01	.01	.05
$\mathbf{s}^{DE}, \{\mathbf{s}^{DE}+1\}$	.78	.26	.16
other: 6 excl.	.01	.34	.36

$r:$	1	2	4
CA	.02	.03	.04
CE	.11	.37	.48
DA	.12	.06	.06
DE	.75	.54	.42

6.5.2 Process with Inequity Aversion

Experiments with finitely repeated public goods games [96] have shown that contribution rates increase by a large magnitude if the possibility for costly individual punishment is provided.[10] At odds with experimental evidence, the standard theory assuming selfish players predicts no punishment in the last period, if punishment is costly. To better account for this evidence Fehr and Schmidt [95] propose a model of inequity aversion, which assumes that players dislike uneven payoffs. The model assumes that people loose utility from being payoff advantaged or disadvantaged, with the loss from an advantage being smaller than the loss from a disadvantage.

Each player is associated with a *coefficient of aversion to disadvantageous inequality* $\alpha_i \geq 0$ and a *coefficient of aversion to advantageous inequality* $\beta_i \geq 0$ satisfying $\beta_i \leq \min\{\alpha_i, 1\}$. If payoffs of players are given by the vector $(x_1, ..., x_n)$, player i receives the (adjusted) utility

[10] That punishment may increase cooperation in social dilemma type games was already observed in earlier experiments [196, 197]. The experiment in [96], however, was the first in which punishment possibility was shown to increase cooperation in the one-shot and finitely repeated public goods games among strangers.

$$u_i^F = x_i - \frac{\alpha_i}{n-1}\sum_j \max\{x_j - x_i, 0\} - \frac{\beta_i}{n-1}\sum_j \max\{x_i - x_j, 0\} \,.$$

Players in the basic process are assumed to be selfish and can be described by having trivial coefficients $\alpha_i = \beta_i = 0$. However, in [95] the distribution of coefficients of inequality aversion is estimated as

$$\alpha_i = \begin{cases} 0 \,;\ pr = 0.3 \\ 0.5 \,;\ pr = 0.3 \\ 1 \,;\ pr = 0.3 \\ 4 \,;\ pr = 0.1 \end{cases} ,\ \beta_i = \begin{cases} 0 \,;\ pr = 0.3 \\ 0.25 \,;\ pr = 0.3 \\ 0.6 \,;\ pr = 0.4 \end{cases} . \tag{6.14}$$

This distribution is shown to be consistent with the data from the experimental public goods game with punishment reported in [96]. In the repeated basic game the exclusion of a defector can be viewed as a form of punishment. In the following we therefore consider an adaptation of the basic process by assuming players whose coefficients of inequity aversion are non-trivial.

We consider a group of n other-regarding players with a uniform scope of forward looking r repeatedly playing the basic game. At each time t each player i plays one of the strategies $\{s_i^{CA}, s_i^{CE}, s_i^{DA}, s_i^{DE}\}$. As in the basic process, at each time $t \geq 2$, before making her move, player i with some fixed probability $q_i \in (0,1)$ updates her own strategy, by observing the vector $\mathbf{s}[t-1]$ of strategies at time $t-1$. Player updates to a strategy maximizing her utility across the next r_i periods, under the assumption that the strategies of the other players do not change in those periods.

Formally, let $\mathbf{s} = (s_1, ..., s_n) \in S$ be the vector of strategies at time $t-1$, and let $\mathbf{s}' = (s_1, ..., s_{i-1}, s_i', s_{i+1}, ..., s_n)$ be the vector of strategies at time t. Player i changes her strategy from s_i to s_i' at t and assumes no other updates in the group during $[t-1, t+r_i]$. Let $x_i(s_i' \mid \mathbf{s})$, defined as in (6.1), be the payoff of player i across the r_i periods following her update, and let $x_{ij}(s_i' \mid \mathbf{s})$, defined as in (6.11), be the payoff to player j across the same periods. The utility of player i for changing her strategy s_i to strategy s_i' is then given by

$$\begin{aligned} u_i^F(s_i' \mid \mathbf{s}) = x_i(s_i' \mid \mathbf{s}) &- \frac{\alpha_i}{n-1}\sum_{j \neq i} \max\{x_{ij}(s_i' \mid \mathbf{s}) - x_i(s_i' \mid \mathbf{s}), 0\} \\ &- \frac{\beta_i}{n-1}\sum_{j \neq i} \max\{x_i(s_i' \mid \mathbf{s}) - x_{ij}(s_i' \mid \mathbf{s}), 0\} \,. \end{aligned} \tag{6.15}$$

If player i is updating at the start of period t she chooses a strategy $s_i[t]$ such that

$$s_i[t] \in \arg\max_{s_i' \in S_i} u_i^F(s_i' \mid \mathbf{s}[t-1]) \; .$$

If this payoff is maximized by the current strategy of the player, the player keeps her current strategy. If not, she randomly selects one of the optimal strategies.

The updating rule outlined above generates a Markov chain with the finite state space S and transition matrix M^F. We refer to this Markov chain the *process with inequity aversion*. The basic process can be seen as a special case of the process with inequity aversion when all coefficients of inequality aversion are trivial. Using the same argument that proved Proposition 6.1 it can be shown that in the process with inequity aversion a state $\mathbf{s}$ is absorbing if and only if

$$s_i \in \arg\max_{s_i' \in S_i} u_i^F(s_i' \mid \mathbf{s}) \; , \tag{6.16}$$

for each player i. Propositions 6.3 and 6.4 hold for any Markov chain and thus for the process with inequity aversion. The proof of the following generalization of Proposition 6.2 for the process with altruism is given in Appendix A.

Proposition 6.7. *Consider the process with inequity aversion and let there be α and β such that $\alpha_i = \alpha$ and $\beta_i = \beta$ for each player i. Let $z \in Z$ be a type distribution. State $\mathbf{s} \in S_z$ is absorbing if and only if each of the states in S_z is absorbing. State $\mathbf{s} \in S_z$ is transient if and only if each of the states in S_z is transient.*

Consequently, whenever players are homogeneous with respect to their coefficients of inequality aversion, we may implement the first class of simulations to detect absorbing and transient states of the process with inequity aversion.

We study two cases with players having homogeneous coefficients of inequality aversion, and one case with players whose coefficients of inequality aversion are distributed according to distribution (6.14). The three processes are characterized by:

F1 $\alpha_i = 40$ and $\beta_i = 1$ for each player i,
F2 $\alpha_i = 4$ and $\beta_i = 0.6$ for each player i, or
F3 for each player i, α_i and β_i are randomly and independently drawn from (6.14).

We assume $r_i = r$ and $q_i = q$ for all i. We consider all the stage game constellations given by Table 6.2 to achieve consistency with parameters considered in the basic process.

As in the model with altruism we conjecture that players with high coefficients of inequality aversion are more likely to achieve cooperation than those with low ones. With this reason we consider the highest feasible coefficients of inequality aversion in F1, and the highest coefficients suggested by (6.14) in F2. Because F1 and F2 consider homogeneous coefficients they may be studied via simulations of first or second class. To study the asymptotic behavior of process F3 we follow the same procedure as outlined in Sec. 6.5.1 for the process A3.

Low Outside Option

Let $o = 0$. For each value of r the dynamics of F1 is shown in the left column and the dynamics of F2 in the right column of Fig. 6.5. In Table 6.7 we report the results of the long run simulations of perturbed processes F1, F2 and F3, respectively. For each of the processes we note the frequencies of the most frequent states and the frequency distribution across different types.

Comparing the dynamics in processes with inequity aversion F1 and F2, shown in the Fig. 6.5, with that of the basic process for $o = 0$, we can see a surprising similarity. The absorbing state $\mathbf{s}^{DA}$ seems to be the unique stable state for every r considered in both F1 and F2, just as it is in the basic process, even though for the process F1 all fully cooperative states are absorbing. These observations are confirmed by comparing the asymptotic behavior of processes F1, F2 and F3, given in Table 6.7, with that given by Table 6.3 for the basic process. The long-run behavior statistics are virtually identical for all these processes, suggesting that state $\mathbf{s}^{DA}$ is the unique stochastically stable state for any distribution of coefficients of inequality aversion in the group. In this sense, we may say that, for $o = 0$, the model of inequality aversion gives the same prediction as the model with selfish players.

High Outside Option

Let $o = 40$. For each value of r the dynamics of F1 is shown in the left column and the dynamics of F2 in the right column of Fig. 6.6. In Table 6.8 we report the results of the long run simulations of perturbed processes F1, F2 and F3, respectively. For each of the processes we note the frequencies of the most frequent states and the frequency distribution across different types.

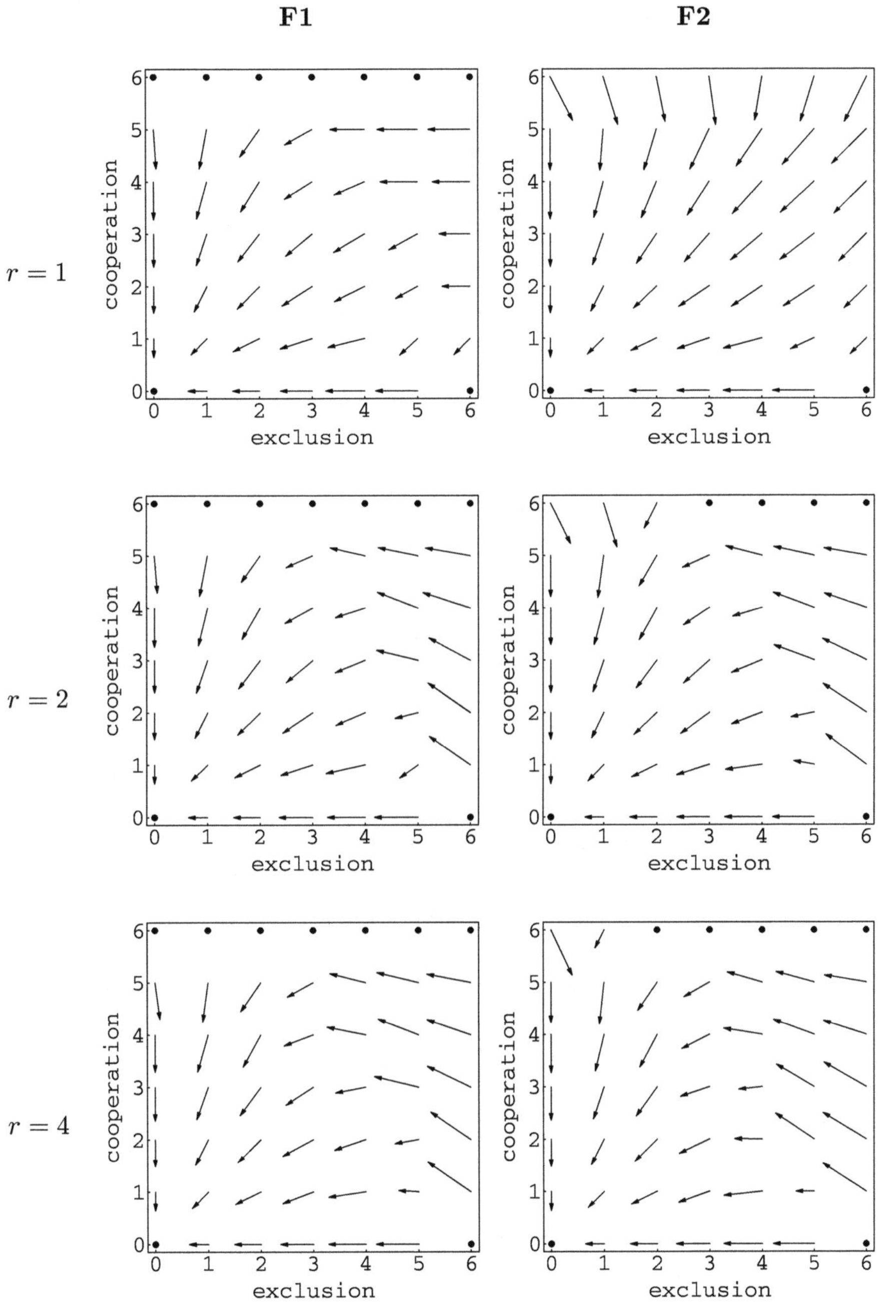

Fig. 6.5. Processes with inequity aversion with o=0. Non-perturbed

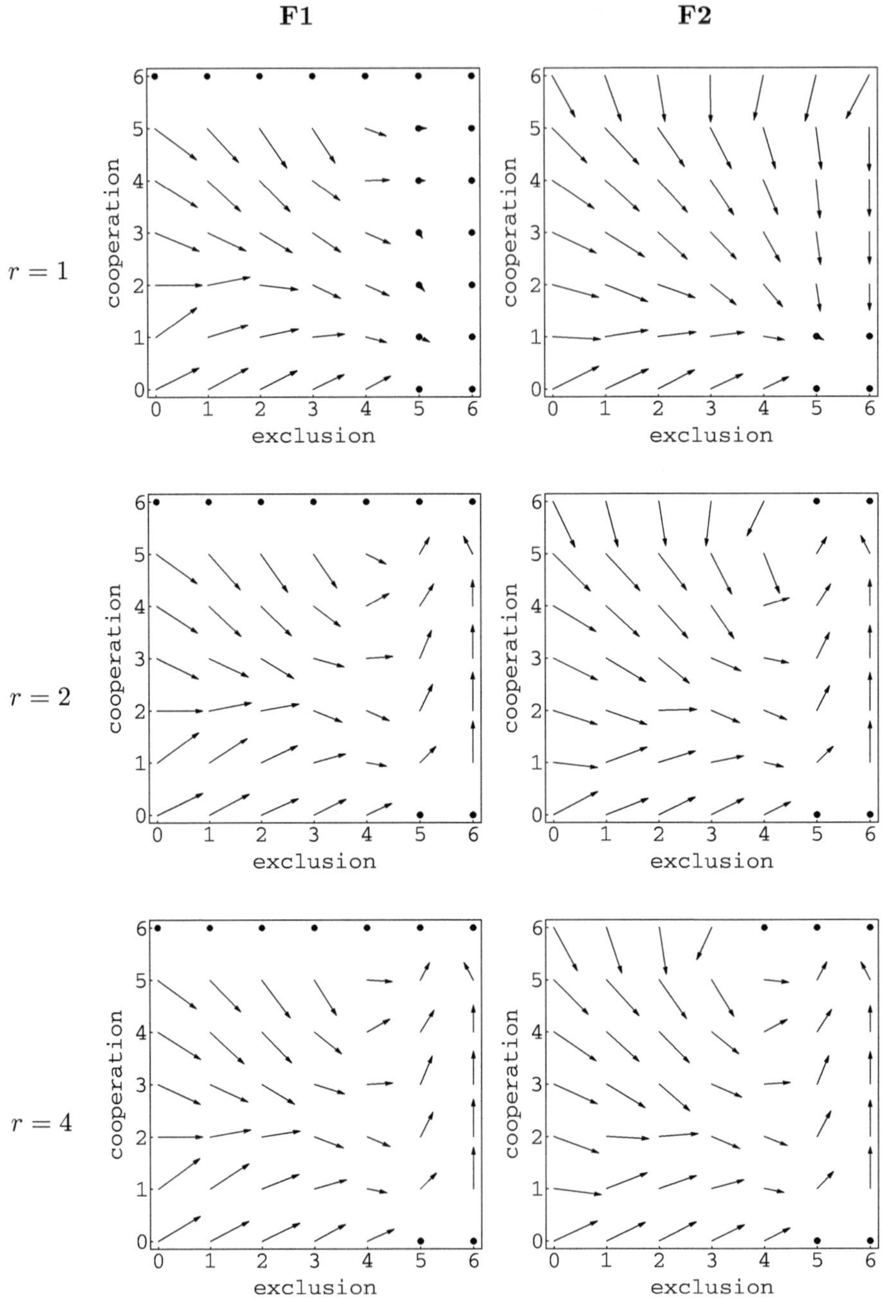

Fig. 6.6. Processes with inequity aversion with o=40. Non-perturbed

Table 6.7. Processes with inequity aversion with o=0. Perturbed

F1

states	$r:$ 1	2	4
6 coop.	.01	.01	.01
s^{DA}	.65	.65	.64
$\{s^{DA}-1\}$	.29	.29	.29

type	$r:$ 1	2	4
CA	.02	.02	.02
CE	.02	.02	.02
DA	.93	.93	.93
DE	.03	.03	.03

F2

states	$r:$ 1	2	4
6 coop.	.01	.01	.01
s^{DA}	.65	.65	.65
$\{s^{DA}-1\}$	.28	.29	.29

type	$r:$ 1	2	4
CA	.02	.02	.02
CE	.02	.02	.02
DA	.93	.93	.93
DE	.03	.03	.03

F3

states	$r:$ 1	2	4
6 coop.	.01	.01	.01
s^{DA}	.65	.65	.64
$\{s^{DA}-1\}$	.28	.29	.29

type	$r:$ 1	2	4
CA	.02	.02	.02
CE	.02	.02	.02
DA	.93	.93	.93
DE	.03	.03	.03

We can compare the results of the simulations of perturbed processes with and without inequity aversion by looking at Tables 6.4 and 6.8. By comparing the frequencies observed in F1, F2, F3 and in the basic process with the high outside option we conclude that cooperation increases with inequity aversion and with r. However, in F2 and F3 cooperation is increased only marginally, and only if players are sufficiently forward looking, that is, if $r \geq 2$.

6.6 Discussion

6.6.1 Results of our Simulation Study

We review and discuss in this section the main results of the chapter. Many conclusions may be drawn simply by examination of the dynamics of the processes, illustrated in Figs. 6.1–6.6. Inference about the long-run behavior of the perturbed processes and about the stability of absorbing states solely on the basis of the figures may, however, be misleading. We therefore analyze and compare the statistics of the

Table 6.8. Processes with inequity aversion with o=40. Perturbed

F1

states	r: 1	2	4
6 coop.	.01	.55	.60
$\mathbf{s}^{DE}, (0,0,1,5)$	.10	.02	.03
other: 6 excl.	.27	.13	.11
other: 5 excl.	.50	.18	.14

type	r: 1	2	4
CA	.02	.10	.12
CE	.32	.78	.76
DA	.12	.04	.04
DE	.54	.08	.08

F2

states	r: 1	2	4
6 coop.	.01	.46	.59
$\mathbf{s}^{DE}, (0,0,1,5),$	.29	.06	.03
other: 6 excl.	.34	.19	.11
other: 5 excl.	.20	.20	.15

type	r: 1	2	4
CA	.02	.07	.11
CE	.11	.69	.75
DA	.12	.05	.04
DE	.75	.19	.10

F3

states	r: 1	2	4
6 coop.	.01	.04	.37
$\mathbf{s}^{DE}, (0,0,1,5),$	.28	.13	.05
other: 6 excl.	.20	.48	.20
other: 5 excl.	.37	.28	.24

type	r: 1	2	4
CA	.02	.03	.05
CE	.11	.47	.71
DA	.12	.06	.06
DE	.75	.44	.18

long-run behavior of the processes, given by Tables 6.3–6.8, but refer to corresponding figures for explanation.

Recall that the process is said to be in a (long-run) equilibrium if it reaches an absorbing state. For reference below we call an equilibrium cooperative if at least one link is established with at least one cooperative player. An absorbing state is called defective if all cooperative players are isolated. The process is in a long-run equilibrium if it reaches a stochastically stable state. When $r = 1$, that is, when players update their strategies according to the myopic best response, we say that their scope of forward looking is trivial.

Result 6.1. OUTSIDE OPTION
A high outside option induces more cooperation than a low outside option.

This result can be explained as follows. When the outside option is $o = 40$ players earn more by excluding defectors than by linking to them. In contrast, with low outside option $o = 0$ exclusion yields no earning, while establishing a link always yields a positive earning, regardless of the actions of linked players. In this sense, exclusion is

Table 6.9. Cooperation in perturbed processes.

$r = 1$

	A1	A2	A3	F1	F2	F3	B
$o = 0$	.94	.04	.04	.04	.04	.04	.04
$o = 40$	.92	.13	.13	.34	.13	.13	.13

$r = 2$

	A1	A2	A3	F1	F2	F3	B
$o = 0$	.94	.37	.04	.04	.04	.04	.04
$o = 40$	.94	.94	.40	.88	.76	.50	.36

$r = 4$

	A1	A2	A3	F1	F2	F3	B
$o = 0$	.94	.37	.05	.04	.04	.04	.04
$o = 40$	.94	.94	.52	.88	.86	.76	.69

costly when the outside option is low and cheap when the outside option is high. Since more exclusion, as a form of retribution, yields more cooperation, more cooperation should be observed with a high outside option.

The above result may be verified by comparing the long-run cooperation levels, for each of the processes considered, between the low and high outside options. Long-run cooperation levels are computed by summing the frequencies of types CA and CE, and are given in Table 6.9. For each particular process and fixed scope of forward looking r the long-run cooperation under $o = 40$ is at least as high as the long-run cooperation under $o = 0$, with the exception of the process A1 with $r = 1$ where the difference is negligible.

Result 6.2. FORWARD LOOKING
Non-trivial forward looking is necessary and sufficient for the existence of cooperative equilibria in the basic process. Under myopic best response no cooperative absorbing states exist. The same is true for processes with altruism or with inequity aversion unless extreme altruism or inequality aversion is assumed.

For the basic process with $o = 0$ the above result is implied by Proposition 6.5. It follows from Fig. 6.2 that, for the basic processes with $o = 40$, there exists a fully cooperative absorbing state when $r = 4$, while all (if any) cooperative players are isolated in any absorbing state when $r = 1$. The absorbing states of the process with altruism A2 are shown in the right-hand columns of Figs. 6.3 and 6.4. The absorbing states of the process with inequity aversion F2 are shown in the right-hand columns of Figs. 6.5 and 6.6. For both processes A2 and F2 and both outside option values considered cooperative absorbing states exist for $r \geq 2$, but not for $r = 1$.

Cooperative equilibria under myopic best response do, however, exist if we assume that the coefficients of altruism or inequity aversion exceed those estimated from the empirical data. The absorbing states of processes A1 and F1 are shown in the left-hand columns of Figs. 6.3–6.6. For both outside option values and all scopes of forward looking considered all fully cooperative states are absorbing.

This suggests that high coefficients of altruism or inequity aversion induce more cooperative equilibria than low ones. Further support for this may be found by comparing the sets of absorbing states between the basic process and the processes with altruism A2 and A1, or between the basic process and the processes with inequity aversion F2 and F1, for fixed o and r. If we do this for all fixed pairs of o and r we may conclude:

Result 6.3. EQUILIBRIA INDUCED BY SOCIAL PREFERENCES
The set of cooperative equilibria of the process with altruism increases with the altruism of players. The set of cooperative equilibria of the process with the inequity aversion increases with inequity aversion of players.

Support for this result is given in Figs. 6.3–6.6. The set of cooperative equilibria of the basic process is contained in that of process A2, which is further contained in that of process A1. Similarly, the set of cooperative equilibria of the basic process is contained in that of process F2, which is further contained in that of process F1.

A large set of cooperative absorbing states, however, does not imply that the updating dynamics are likely to direct the process toward these states. A case in point is given by Fig. 6.1 which illustrates the dynamics of the basic process with $o = 0$. It is evident from the figure that, while increasing the scope of forward looking r increases the set of cooperative absorbing states, the dynamics remain mostly unaffected and direct the basic process toward the defective absorbing state $\mathbf{s}^{DA}$ for all considered r. Table 6.3 confirms that the perturbed basic process with $o = 0$ spends most of the time in the state $\mathbf{s}^{DA}$. This indicates that $\mathbf{s}^{DA}$ is the unique stochastically stable state of this process.

On the other hand, even if all equilibria are defective, cooperation may be substantial in the presence of persistent errors. The basic process with $o = 40$ and $r = 2$ is one such example. It is evident from Fig.

6.2(b) and Table 6.4 that no cooperative absorbing states of the non-perturbed process exist, but yet the slight perturbation of the process leads to a substantial level of cooperation.

The following two results summarize the outcomes of the long-run simulations of perturbed processes.

Result 6.4. LONG-RUN EQUILIBRIA, LOW OUTSIDE OPTION
Consider $o = 0$. *State* $\mathbf{s}^{DA}$, *in which all players defect and form a complete network, is the unique stochastically stable state of the basic process, as well as of the process with inequity aversion, for any distribution of inequity aversion and for all considered* r. *It is also the unique stochastically stable state of the process with altruism when, as suggested by the model of altruism, the distribution of altruism is given by (6.10). Cooperative absorbing states may be stochastically stable only when all players have (extremely) high coefficients of altruism, such as in processes with altruism A1 and A2.*

Result 6.5. LONG-RUN EQUILIBRIA, HIGH OUTSIDE OPTION
Consider $o = 40$. *When* $r \leq 2$ *there are no cooperative stochastically stable states of the basic process. When* $r = 2$, *however, a small perturbation of the basic process may lead the dynamics away from defective states and, consequently, medium levels of cooperation may be achieved. When* $r = 4$ *the state* $\mathbf{s}^{CE}$, *in which all players cooperate and exclude defectors, is the unique stochastically stable state of the basic process. If, as suggested by the corresponding theories, coefficients of altruism or inequity aversion are drawn randomly from the distributions (6.10) or (6.14) the processes with altruism and inequity aversion exhibit the same dynamics as the basic process. Cooperation, however, increases when these coefficients are (extremely) high, such as in processes A1, A2, F1, or F2.*

The stochastically stable states and the long-run behavior of each process may be inferred from the frequencies of visits of different absorbing states by the process, given in the left-hand columns of Tables 6.3–6.8. The final result summarizes the comparison of the basic process with processes with altruism and inequity aversion from the discussion above.

Result 6.6. COOPERATION INDUCED BY SOCIAL PREFERENCES
If coefficients of altruism are distributed by (6.10), as suggested by the model of altruism, the process of altruism exhibits qualitatively the same behavior as the basic process assuming selfish players. Similarly, if coefficients of inequity aversion are distributed by (6.14), as suggested by

the model of inequality aversion, the process of inequity aversion exhibits qualitatively the same behavior as the basic process. For $o = 0$ *the process of inequity aversion exhibits qualitatively the same behavior as the basic process for any distribution of coefficients of inequity aversion.*

In Chap. 5 we discuss the dynamics of behavior of human subjects playing the finitely repeated basic game with the low outside option in a laboratory experiment. The game was repeated for 60 periods. The behavior in the experiment differed significantly between the last 5 periods and the periods 1-55, which is in accordance with the assumption, made in this chapter, that people look a few periods ahead. High average levels of cooperation were observed during most of the periods 1-55, while a strong tendency toward defection was evident in the final 5 periods, the effect known in experimental repeated games as the "end-game effect".

To see that this is consistent with the limited forward looking assumption, consider players whose scope of forward looking r is sufficiently long to induce cooperation during initial periods. When the play approaches the final periods of the repeated game players adjust their behavior to the shorter scope of the future. A player, for example, who realizes that there are only $\tau < r$ periods left may behave as if having a shorter scope of forward looking τ. For sufficiently short τ players may thus defect, which would lead to the last period decline in cooperation.

Among the processes with low outside option considered in this chapter only the process with altruism A2 reaches substantial cooperation with $r \geq 2$ and low cooperation with $r = 1$ (see Table 6.9). This suggests both (the possibility of) high altruism as well as the non-trivial forward looking as determinants of behavior in our experiments.

6.6.2 Discussion of the Model Assumptions

Some caution may be in order in the interpretation of our results. Firstly, our set of strategies was restricted by assumptions of limited memory and consistency. Enlarging the set of strategies may affect the results. However, it is not obvious which set of strategies would best represent those used in real life. One way to circumvent this problem would be to allow for all feasible strategies via the methodology of genetic algorithms (this approach is endorsed in [188]). The nature of this approach, however, means that the evolution of strategies is not easily tractable and the interpretation of (success of) the complex surviving strategies is left to speculation [48]. Redundant sophistication yielded by genetic algorithms may sometimes be eliminated by adding the cost of complexity, e.g. as in [198]. The manner in which cost is related to complexity may, however, bias the evolution. Compare, for example, [199] and [200].

Another approach is to consider relaxed assumptions about the memory length and consistency of strategies. However, the relaxed bounds are just as artificial as any other, while the resulting set of all possible strategies soon becomes too large to permit straightforward analysis or interpretation. An occasional compromise is to consider a large set of intuitive strategies inferred from real life or experimental evidence [47]. A set of strategies obtained in this way is, however, far from complete within the considered boundaries, as well as restricted by a subjective judgment.

Best response and imitation models usually assume that players know the strategies played by the other players. This, however, becomes a strong assumption if players use relatively complex strategies. It may be reasonable to assume that actions are perfectly observable, but they need not reveal the strategies that yield these actions. In a world of complex strategies the dynamics of players' beliefs about strategies of the other players could therefore be better described by a model of learning from observed actions such as Bayesian learning [201, 202, 203] (see [204] for overviews and tests of models of learning in games).

A model of adaptive play that may allow for complex strategies without such strong assumptions is that of reinforcement learning [205]. In the future we may complement our results with a reinforcement learning model that allows limited forward looking. In this chapter, however, we opt for a best response model within a small but complete set of strategies within clearly defined and intuitive restrictions. The resulting dynamics appears complex yet sufficiently tractable to allow straightforward interpretation.

The second caution, common to all simulation studies, concerns the sensitivity of results to the specific choice of parameters in the simulations. In particular, we considered a single set of prisoner's dilemma game parameters, group size $n = 6$, updating probability $q = 0.2$, and error probability $\varepsilon = 0.1$. We next discuss how changing these parameters may affect our results.

The error probability does affect the process dynamics, and its impact was described in Sec. 6.3.2. In particular, Theorem 6.1 describes how the asymptotic behavior of the process changes as ε decreases toward zero. It states that for sufficiently small ε the process is most likely to be observed in one of its stochastically stable states. We, however, have shown that if perturbation is small but not negligible the long-run behavior of the process may differ greatly from this limiting prediction. One such example is given by Result 6.5 for our basic process with $o = 40$ and $r = 2$.

The impact of the prisoner's dilemma game parameters for the process dynamics requires more thorough elaboration. For this we perform a test of the robustness of our main results to a change in the value of $f = v(D, C)$, the prisoner's dilemma game payoff when defecting on a cooperator. This value may be seen as a proxy for the incentive to defect. We use the simulations of the second class to detect the asymptotic behavior of the basic process, the process with altruism A3, and the process with inequity aversion F3, for 26 different values of f, equidistantly taken from the interval $[50, 100]$. We assume $r = 2$ in all these simulations. For each value of f and each of the two outside options considered we register the average numbers of cooperative players in each of the three processes. Figure 6.7 shows the graphs of average cooperation depending on the value of f, for each of the processes and outside option values. For low outside option cooperation levels remain negligible for all processes and values of f. For high outside option, however, the value of f does have an effect on cooperation. It is interesting to note that for low f, close to $c = 50$, cooperation is lower in the process with altruism A3 than in the basic process. This is most likely due to the presence of spiteful players with negative coefficients of altruism, which gain utility by decreasing the payoff of other players.

Finally, our definition of forward looking is a relatively simple way of modeling the relation between one's own present behavior and the future behavior of the others. We assume that players believe, when they are updating their own strategy, that no other players will update in the periods covered by their scope of forward looking. This assumption seems a fair approximation of the (short span) future behavior if, as is the case in our model, the possibilities for updates are rare, and it is correct once the play reaches a long-run equilibrium. It would nevertheless be worthwhile to relax it in a model that allows for players who understand that (and how) other players may revise their strategies in the future. Such players might, for example, strictly prefer exclusion from inclusion, even when it is costly to exclude, because they realize that threatening with exclusion induces other players to abandon defective strategies.

Jehiel [170] suggested such a model when introducing the concept of limited foresight. Players with limited foresight in each period choose

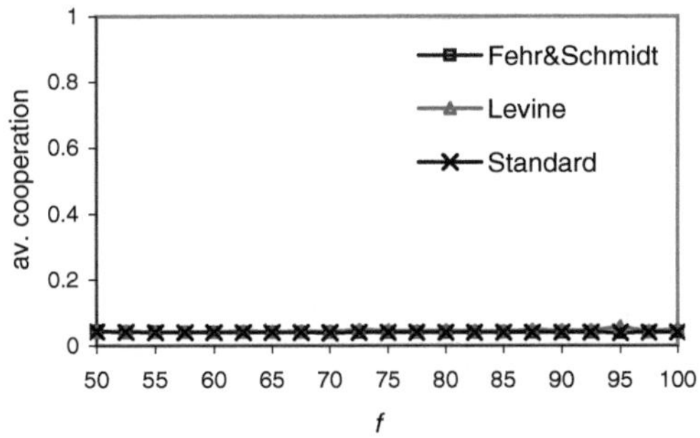

a) $o = 0$

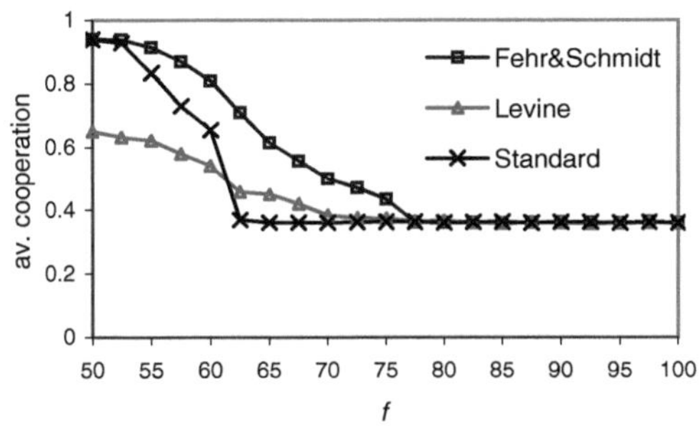

b) $o = 40$

Fig. 6.7. Average cooperation in simulations of the basic process, the process with altruism A3, and the process with fairness F3, for different values of f, with $r = 2$ and (**a**) low outside option $o = 0$; (**b**) high outside option $o = 40$. Note that $f = 70$ is chosen in all simulations in the previous sections

an action which maximizes their payoff across a finite number of coming periods, believing that play is going to follow a particular sequence of action profiles. The game is said to be in a limited foresight equilibrium if, at each period, beliefs of all players coincide with the actual path of play. In [206] a model of learning about the strategy used by the opponent is developed, that leads to a limited forecast equilibrium in two-player games. Such an equilibrium considers players who at each period use the truncated history of play to update their belief about the future paths of play. Adjusting this concept to the setting of our n-player stage game may, however, prove difficult, among other things due to the large number of possible action paths.

6.7 Conclusion

In this chapter we study the dynamics of partner choice and cooperation among boundedly rational players that to a limited extent understand the consequences of their own actions and take them into account when adapting their behavior. We show that the assumption of limited forward looking induces cooperative equilibria even if players are best responding to their neighborhood. These equilibria, however, are often not stable with respect to occasional deviations from the best response behavior that we assume players make. In particular, when exclusion is costly cooperation remains low. It turns out that limited forward looking increases cooperation only when exclusion is sufficiently cheap.

Assuming that players are other-regarding, rather than selfish, does not substantially change these results. We consider in this chapter two theories of social preferences: the model of altruism and the model of inequality aversion. Assuming altruistic or inequity averse players further increases the set of cooperative equilibria. However, if altruism or inequity aversion is distributed according to the distributions proposed by the corresponding models the dynamics of behavior among other-regarding players does qualitatively not differ from that among selfish players. This suggests that, at least in the setting of the repeated prisoner's dilemma game played on an endogenous network, a limited comprehension of the reactions to one's own actions may be as important for sustaining cooperation as is a concern for others.

A

Appendix: Proofs

A.1 Proofs from Chapter 4

For convenience we introduce the indexing function $\overleftrightarrow{(\cdot)} : \{-n+1, ..., 2n\} \rightarrow \{1, 2, ..., n\}$, defined by

$$\overleftrightarrow{i} = \begin{cases} i-n & \text{if } i \in \{n+1, ..., 2n\} \\ i & \text{if } i \in \{1, ..., n\} \\ i+n & \text{if } i \in \{-n+1, ..., 0\} \end{cases} . \tag{A.1}$$

The application of this function is as follows. Place players on the circle such that player i follows player $i-1$ and that player 1 follows player n, clockwise. For any i the indices of her m immediate neighbors, going in one direction along the circle, are given by $\overleftrightarrow{i-1}, ..., \overleftrightarrow{i-m}$, and by $\overleftrightarrow{i+1}, ..., \overleftrightarrow{i+m}$, going in the other direction. For even n the players i and $\overleftrightarrow{i+n/2}$ are directly opposite on the circle.

The proof of Proposition 4.8 relies on the following lemmas.

Lemma A.1. *Take k such that $2 \leq k < n$. Consider $\Gamma(\mathbf{k})$ with $d > 0$ and $k_i = k$ for each i.*

a. If k and n are odd, there exists a linking-proof network g such that $l_j(g) = k-1$ for some j and $l_i(g) = k$ for all $i \neq j$.
b. Otherwise, there exists a linking-proof network g such that $l_i(g) = k$ for all players i.

Proof. (b.1) If k is even then g is constructed by linking each player i to players $\overleftrightarrow{i-1}, ..., \overleftrightarrow{i-k/2}$ and $\overleftrightarrow{i+1}, ..., \overleftrightarrow{i+k/2}$. All players thus establish k links, hence network is linking-proof.

(b.2) If k is odd but n is even then g is constructed by linking each player i to players $\overleftrightarrow{i-1}, ..., \overleftrightarrow{i-(k-1)/2}$ and $\overleftrightarrow{i+1}, ..., \overleftrightarrow{i+(k-1)/2}$, and to player $\overleftrightarrow{i+n/2}$. All players thus establish k links, hence network is linking-proof.

(a) If k and n are both odd then a network g' is constructed by linking each player i to players $\overleftrightarrow{i-1}, ..., \overleftrightarrow{i-(k-1)/2}$ and players $\overleftrightarrow{i+1}, ..., \overleftrightarrow{i+(k-1)/2}$. All players thus establish $k-1$ links. Network g is constructed from g' by linking each of the $(n-1)/2$ disjoint pairs of separated players$\{(i,(n-1)/2+i)\}_{i=1}^{(n-1)/2}$. Player n was not assigned to any pair and thus establishes $k-1$ links in g, while all other players establish k links, hence network g is linking-proof. □

Lemma A.2. *Consider $\Gamma(\mathbf{k})$ with $d > 0$ and $2 \leq k_i < \sqrt{2n-7}$ for each i. Take any stable-equilibrium network $g \in \mathcal{S}(\Gamma(\mathbf{k}))$. For any pair of players i and j there exists another pair of players i' and j' such that $g_{ii'} = 0, g_{jj'} = 0$, and $g_{i'j'} = 1$.*

Proof. The proof proceeds as follows. Let $k = \max_{i \in N} k_i$. Fix i and j. We characterize the set of players $\overline{M_{ij}}$ that cannot be in the role of i'or j'. We show that the size of this set depends on k. We characterize this dependence and prove that the size of $\overline{M_{ij}}$ is less or equal to $n-2$ if $k < \sqrt{2n-7}$. If the size of $\overline{M_{ij}}$ is less or equal to $n-2$ then there must be two players $i'$and j' satisfying the conditions of the Lemma.

We say that a pair of linked players i' and j' complement ij if $g_{ii'} = 0$ and $g_{jj'} = 0$. If xy is not separated from ij it must be that either (a) $g_{ix} = g_{iy} = 1$, or (b) $g_{jx} = g_{jy} = 1$, or (c) $g_{ix} = g_{jx} = 1$, or (d) $g_{iy} = g_{jy} = 1$, that is, either one of i, j is linked to both x, y, or both i, j are linked to one of x, y. Define sets O_{ij}, N_{ij}, N_i, and N_j as follows: $N_{ij} = L_i(g) \cap L_j(g)$, $N_i = L_i(g) \backslash N_{ij}$, $N_j = L_j(g) \backslash N_{ij}$, and $O_{ij} = N \backslash (L_i(g) \cup L_j(g))$. N_{ij} is the set of players linked to both i and j, N_i and N_j are sets of players linked to one of i, j, and O_{ij} is the set of remaining players, separated from both i and j. Let $M_{ij} = \{i, j\} \cup L_i(g) \cup L_j(g)$.

If a pair of linked players x, y does not complement ij, then either (i) both of them belong to N_i or both to N_j, or (ii) at least one of them belongs to N_{ij}. Let

$$\overline{N_{ij}} = \{x \mid y \in N_{ij} \text{ for each } y \text{ s.t. } g_{xy} = 1\} \backslash M_{ij}$$

be the set of players, separated from i and j, but whose neighbors all belong to N_{ij}. Let $\overline{M_{ij}} = M_{ij} \cup \overline{N_{ij}}$ be the set of players whose neighbors are either i or j or belong all to N_{ij}, including players i and j. If a pair of neighbors x, y, $g_{xy} = 1$ does not complement ij, both x and y belong to $\overline{M_{ij}}$.

All but at most one player in $\overline{N_{ij}}$ establish their maximal number of links in g. To see this assume $x, y \in \overline{N_{ij}}$ such that $l_x(g) < k_x$ and $l_y(g) < k_y$. Because g is a linking-proof network, Proposition 4.3 implies that $g_{xy} = 1$. Then, by definition of $\overline{N_{ij}}$, both x, y should belong to N_{ij}. This, however, is not possible, since they both belong to $\overline{N_{ij}}$.

Since $k_x \geq 2$ for all $x \in \overline{N_{ij}}$ and since at most one of them does not establish all her links, all but at most one of them is linked to at least two players in N_{ij}. This restricts the maximal possible number of players in N_{ij}. Let $n_{ij} = |N_{ij}|$, $n_i = |N_i|$, and $n_j = |N_j|$. Each player in N_{ij} is already linked to two players, i and j, and can have at most $k - 2$ more links. Number n_{ij} is maximized if each player in N_{ij} makes $k - 2$ links with players in $\overline{N_{ij}}$, each of whom makes two links, apart from one who makes one link. The joint number of links made by players in N_{ij} is thus $n_{ij}(k - 2)$ and each player in $\overline{N_{ij}}$ shares two of these, aside from at most one who shares only one link. The number of players in $\overline{N_{ij}}$ is therefore bounded above by

$$\left|\overline{N_{ij}}\right| \leq (n_{ij}(k-2) - 1)/2 + 1 \ .$$

Note that $\overline{M_{ij}} = \overline{N_{ij}} \cup N_{ij} \cup N_i \cup N_j \cup \{i, j\}$ with no overlap between pairs of sets. This implies

$$\left|\overline{M_{ij}}\right| \leq (n_{ij}(k-2)-1)/2+1+n_{ij}+n_i+n_j+2 \leq (n_{ij}(k-2)-1)/2+3+k \ ,$$

since $n_{ij} + n_i + n_j \leq k$. This bound is maximized when $n_{ij} = k$, in which case it becomes

$$\left|\overline{M_{ij}}\right| \leq (k^2 + 5)/2$$

By assumption, $k < \sqrt{2n-7}$, which implies that $\left|\overline{M_{ij}}\right| \leq n - 2$. At least two players do not belong to $\overline{M_{ij}}$. Following Proposition 4.3 at least one of them makes at least one link. There must then be a pair of neighbors which complement ij. □

Lemma A.3. *Consider $\Gamma(\mathbf{k})$ with $d > 0$ and $2 \leq k_i < \sqrt{2n-7}$ for each i. There exists a linking-proof network g such that $2 \leq l_j(g) \leq k_j$ for some player j and $l_i(g) = k_i$ for all other players $i \neq j$.*

Proof. The proof proceeds constructively as follows. We create, by an iterative procedure, an initial linking-proof network (LPN) g^0 in which each player establishes at least two links. In this network there may be several players that miss one or more links. We then use Lemma A.2 iteratively to create a finite sequence of LPN $g^1, ..., g^\Phi$ such that the total number of links strictly increases. At most one player is missing links in g^{Φ}.

Let g^0 be the network generated by the following procedure:

1. establish links between each pair $\{i, i+1\}$ and between the pair $\{n, 1\}$.
2. repeat: if there exists a pair of separated players i and j such that $l_i < k_i$, $l_j < k_j$, establish the link ij; until no such pair exists.

Because the number of established links in the above procedure strictly increases, but the maximal possible number of links is finite, the procedure stops in a finite number of steps. By Proposition 4.3, the network g^0 is LPN as there exists no pair of separated players i and j such that $l_i(g^0) < k_i$, $l_j(g^0) < k_j$.

Let networks $g^1, g^2, ...$ be generated by the following procedure for $\phi = 0, 1, ...$:

As long as in g^{ϕ} there exist two players i and j such that $l_i(g^{\phi}) < k_i$, $l_j(g^{\phi}) < k_j$, then

1. take a pair of neighbors $i'j'$, $g^{\phi}_{i'j'} = 1$ which complement ij, that is, $g^{\phi}_{ii'} = 0$ and $g^{\phi}_{jj'} = 0$ (we show in Lemma A.2 that such pair exists),
2. let $g^{\phi+1}$ coincide in all links with g^{ϕ}, aside from $g^{\phi+1}_{ii'} = 1$, $g^{\phi+1}_{jj'} = 1$, and $g^{\phi+1}_{i'j'} = 0$.

The procedure ends when at most one player is still missing links. Again, the number of established links strictly increases with steps and the procedure stops in finite number Φ of steps. In g^{Φ} all players establish their maximal number of links, aside from at most one player who has at least two links.

To see that each $g^{\phi+1}$ is LPN, proceed again iteratively. Let g^{ϕ} be LPN and satisfy the linking constraints. Let ij and $i'j'$ be the two pairs of players chosen by the above procedure in the corresponding step. Then, $g^{\phi+1}$ also satisfies linking constraints. Furthermore, by Proposition 4.3, $g^{\phi}_{ij} = 1$, and thus $g^{\phi+1}_{ij} = 1$. Also by Proposition 4.3, because g^{ϕ} is LPN and $l_i(g^{\phi}) < k_i$ and $g^{\phi}_{ii'} = 0$, then $l_{i'}(g^{\phi}) = k_{i'}$. Similarly, $l_{j'}(g^{\phi}) = k_{j'}$. The only link that is removed is $i'j'$, but as players i' and j' establish the same number of links in both g^{ϕ} and $g^{\phi+1}$, the $g^{\phi+1}$ is a LPN. Hence, network g^{Φ} is also LPN. □

Proof of Proposition 4.8. (1) Assume that $k_i = k$ for each i. Let $2 \leq k \leq n-2$. In the following we apply Lemma A.1.

(a) Let k be even. For every i the remaining players $N \backslash \{i\}$ can form a linking-proof network (LPN) g^i such that $l_j(g^i) = k$ for each $j \neq i$

and $l_i(g^i) = 0$. Furthermore, a LPN g can be established such that $l_i(g) = k$ for each i, see Lemma A.1 (b). (b) Let k be odd and n even. For every i the remaining $n-1$ players $N\backslash\{i\}$ can form a LPN g^i such that $l_j(g^i) = k$ for each $j \neq i$ and $l_i(g^i) = 1$, see Lemma A.1 (a). A LPN g can also be established such that $l_i(g) = k$ for each i. (c) Let k and n both be odd. For every i the remaining $n-1$ players $N\backslash\{i\}$ can form a LPN g^i such that $l_j(g^i) = k$ for each $j \neq i$ and $l_i(g^i) = 0$. A LPN g can also be established such that $l_i(g) \geq k-1$ for each i.

In all three cases (a), (b), and (c), $l_i(g^i) < l_i(g)$ for each i. This fulfills the conditions of Proposition 4.7.

(2) Let $2 \leq k_i < \sqrt{2n-9} = \sqrt{2(n-1)-7}$ for each i. According to Lemma A.3, for any subgroup of players $N\backslash\{i\}$ of size $(n-1)$ there exists a network $g^{N\backslash i}$ such that $l_j(g^{N\backslash i}) \leq k_j$ for some $j \in N\backslash\{i\}$ and $l_x(g^{N\backslash i}) = k_x$ for all other $x \in N\backslash\{i,j\}$. Let g^i be the network among players N which is obtained from $g^{N\backslash i}$ by adding player i and (i) adding the link ij if $l_j(g^{N\backslash i}) < k_j$ or (ii) adding no links if $l_j(g^{N\backslash i}) = k_j$. Aside from player i who establishes at most one link, all other players establish their maximal number of links in g^i, hence, following Proposition 4.3, g^i is a linking-proof network (LPN).

We have shown that for each player i there exists a LPN g^i such that $l_i(g^i) \leq 1$. Furthermore, according to Lemma A.3, there also exists a network g such that $l_i(g) \geq 2$ for each i. The conditions of Proposition 4.7 are thus satisfied. □

Proof of Theorem 4.1. All threats of the recursive trigger profile are credible, because they consist of a repetition of a static equilibrium. We thus need to verify that there exist positive integers $t^0, ..., t^m$ such that all threats are effective.

A threat against an early defection of player $i \in N^0$ is effective if the one-period profit from the defection is offset by the loss incurred through subsequent exclusion. A threat against an early deviation of player $i \notin N^0$ in period $t \in \{1, ..., T - t^{\eta(i)}\}$ is effective if the profit from the deviation collected during $\{t, ..., T - t^{\eta(i)}\}$ is offset by the loss of reward during $\{T - t^{\eta(i)} + 1, ..., T - t^{\eta(i)-1}\}$. By an early defection a player may avoid that other players free-ride on her, and may thus gain for more than one period. Namely, if some player defects early then all players defect in all subsequent periods. Player $i \in N^{\eta(i)}$ may thus avoid being free-rided upon by her neighbors from $N^{\eta(i)+1}$ during $\{T - t^{\eta(i)+1} + 1, ..., T - t^{\eta(i)}\}$. We need to take this potential profit into account to determine the effectiveness of threats.

Let γ_i be the per-period difference in payoff to player i for avoiding being free-ridden upon during $\{T - t^{\eta(i)+1} + 1, ..., T - t^{\eta(i)}\}$. This value depends on the number of neighbors of i in sets $\underline{L_i} = L_i \cap (N^{\eta(i)} \cup N^{\eta(i)-1})$ and $\overline{L_i} = L_i \cap N^{\eta(i)+1}$. If there were no deviations then all players in $\underline{L_i}$ cooperate and all players in $\overline{L_i}$ defect during the above mentioned periods. However, if i deviated early then all players defect during these periods. Player i thus looses because her neighbors in $\underline{L_i}$ do not cooperate, and gains because her neighbors in $\underline{L_i}$ do not free-ride on her. Hence, γ_i can be positive, negative or 0. Let $\overline{\gamma}_i = \max\{0, \gamma_i\}$.

Let α_i be the maximal potential one-period profit to player i from an early defection. Let β_i be the per-period loss to player i from an exercised threat in any of the periods $\{2, ..., T - t^{\eta(i)}\}$ (the first deviation can take place in period 1 and the first threat is then exercised in period 2). Finally, let δ_i be the per-period loss to player i when foregoing the reward of free-riding during $\{T - t^{\eta(i)} + 1, ..., T - t^{\eta(i)-1}\}$, or when $i \in N^0$ is punished by exclusion during $\{T - t^0 + 1, ..., T\}$. The following holds for each i: $\alpha_i, \beta_i, \delta_i > 0$, while $\overline{\gamma}_i \geq 0$.

Define $\Delta_0 = t^0$ and $\Delta_\eta = t^\eta - t^{\eta-1}$ for $\eta = 1, ..., m$. Let $\Lambda(i,t)$ be the difference in the total payoff to player i for an early defection in period $t \in \{1, .., T - t^{\eta(i)}\}$. Then,

$$\begin{aligned}\Lambda(i,t) &\leq \alpha_i - (T - t^{\eta(i)+1} - 1)\beta_i + \Delta_{\eta+1}\overline{\gamma}_i - \Delta_\eta \delta_i \\ &\leq \alpha_i + \Delta_{\eta+1}\overline{\gamma}_i - \Delta_\eta \delta_i .\end{aligned}$$

All threats are surely effective if $\alpha_i + \Delta_{\eta+1}\overline{\gamma}_i - \Delta_\eta \delta_i < 0$ for each i.

Players in N^m do not award any player by letting her free ride. Hence, $\overline{\gamma}_i = 0$ for $i \in N^m$. The threats for these players are thus effective if $\alpha_i - \Delta_m \delta_i < 0$, which rewrites as

$$\Delta_m > \max_{i \in N^m} \frac{\alpha_i}{\delta_i} . \tag{A.2}$$

There exists a finite positive integer Δ_m which satisfies (A.2). Similarly, for $\eta \in \{0, ..., m-1\}$, the threats for players in N^η are effective if

$$\Delta_\eta > \max_{i \in N^\eta} \frac{\alpha_i + \Delta_{\eta+1}\overline{\gamma}_i}{\delta_i} . \tag{A.3}$$

Given a finite $\Delta_{\eta+1}$, there exists a finite positive integer Δ_η which satisfies (A.3). Starting with Δ_m it is now possible to recursively determine the minimal Δ_η such that (A.2) and (A.3) are satisfied for all η. Setting $t^\eta = \sum_{\zeta=0}^{\eta} \Delta_\zeta$ we get positive integers $t^0, ..., t^m$ such that all threats are effective. Setting $T^* = t^m$ we conclude that $\varrho(g^*, g^*, \{g^i\}_{i \in N^0}, T, (t^\eta)_{\eta=0}^m)$ is subgame perfect for any $T \geq T^* + 1$. $\square$

Proof of Proposition 4.10. Consider the following construction of a connected LPN. Begin with a wheel network connecting all players. Now recursively add links between pairs of separated players who can still add links. By Proposition 4.3 the network is LPN when there is no such pair of players. This construction stops in finite time as there is a finite number of possible links.

The complete network is not feasible because $k_j \leq n-2$. Let g^* be a connected LPN. There exist separated players i' and i''. There also exists a shortest path $(i' = i^0, i^1, ..., i^\kappa = i'')$ between each such pair. Define a partition of N into players M^1 who establish all their links and players M^0 who do not:

$$M^0 = \{i \in N \mid l_i(g^*) < k_i\} \text{ and}$$
$$M^1 = \{i \in N \mid l_i(g^*) = k_i\}.$$

By Proposition 4.3 all players in M^0 are linked. Set M^1 cannot be empty as otherwise all players would be linked and the network complete. In the rest of the proof we consider all possible cases of how pairs of separated players are distributed between M^0 and M^1. For each case we construct another LPN in which one player has less neighbors than in g^*.

a) Assume that M^0 is empty. Take separated players i' and i'' and a shortest path $(i' = i^0, i^1, ..., i^\kappa = i'')$. A network g^{i^1} can be constructed from g^* by removing links $i^0 i^1$ and $i^1 i^2$ and adding the link $i^0 i^2$. By Proposition 4.3 g^{i^1} is LPN because i^1 is the only player who does not establish all her links.

b) Assume that both M^0 and M^1 are non-empty. Assume also that $g^*_{ii'} = 1$ for all $i \in M^0$ and $i' \in M^1$. Then there exist separated players $i^0, i^2 \in M^1$ and a shortest path (i^0, i^1, i^2) with $i^1 \in M^0$. A network g^{i^1} can be constructed from g^* by removing links $i^0 i^1$ and $i^1 i^2$ and adding the link $i^0 i^2$. The sets M^0 and M^1 thus remain the same and no links between players in M^0 change. In g^{i^1} all players in M^0 are still linked. By Proposition 4.3 g^{i^1} is LPN.

c) Assume now that both M^0 and M^1 are non-empty and take separated players $i \in M^0$ and $i' \in M^1$ with a shortest path $(i = i^0, i^1, ..., i^\kappa = i')$. Because players i and i^2 are not linked, but all players in M^0 must be linked, it must be that $i^2 \in M^1$. W.l.o.g. let $i^2 = i'$.

c.1) Assume that $i^1 \in M^0$. Player i^0 has no link to i^2 and can add links to those she has in g^*. A network g^{i^1} can be constructed from g^* by removing link $i^1 i^2$ and adding the link $i^0 i^2$. The new set $M^0(g^{i^1})$ is a subset of M^0 and no links between players in M^0 change. All players in $M^0(g^{i^1})$ are still linked. By Proposition 4.3 g^{i^1} is LPN.

c.2) Assume that $i^1 \in M^1$ and that $g^*_{i''i^1} = 1$ for all $i'' \in M^0$. Hence, all players in set $M^0 \cup \{i^1\}$ are linked. Again, a network g^{i^1} can be constructed from g^* by removing the link i^1i^2 and adding the link i^0i^2. The new set $M^0(g^{i^1})$ is a subset of $M^0 \cup \{i^1\}$ and no links between players in $M^0 \cup \{i^1\}$ change. All players in $M^0(g^{i^1})$ are linked. By Proposition 4.3 g^{i^1} is LPN.

c.3) Finally, assume that $i^1 \in M^1$ and let $g^*_{i''i^1} = 0$ for some $i'' \in M^0$. Set $j^0 = i''$, $j^1 = i$ and $j^2 = i^1$ and construct a LPN g^{j^1} as in (c.1).

In cases a) - c.2) above we constructed a LPN g^{i^1} such that $l_{i^1}(g^{i^1}) < l_{i^1}(g^*)$. In (c.2) we constructed a LPN g^{j^1} such that $l_{j^1}(g^{j^1}) < l_{i^1}(g^*)$. In all cases above the conditions of Theorem 4.1 are thus satisfied. □

A.2 Proofs from Chapter 5

Proof of Proposition 5.1. For each player i and for any profile of actions of her opponents a^*_{-i}: compared to defection, cooperation decreases her direct value by 20 per neighbor and increases the direct values of each neighbor by 40. The direct values of the remaining players are not affected. Hence, the difference in the utility between cooperation and defection, $\Delta^C u_i \equiv u_i^A(C, a^*_{-i}, p^*) - u_i^A(D, a^*_{-i}, p^*)$, is

$$\begin{aligned}
\Delta^C u_i &= -20\; l_i(g^*) + 40 \sum_{j \in L_i(g)} \frac{\gamma_i + \lambda \gamma_j}{1+\lambda} \\
&= 40\; l_i(g^*) \frac{\gamma_i + \lambda \widehat{\gamma}_{-i}(g^*)}{1+\lambda} - 20\; l_i(g^*) \\
&= 20\; l_i(g^*) \left(2 \frac{\gamma_i + \lambda \widehat{\gamma}_{-i}(g^*)}{1+\lambda} - 1 \right) .
\end{aligned}$$

In the Nash equilibrium, if $\Delta^C u_i > 0$ then it must be that player i cooperates. Alternatively, if $\Delta^C u_i < 0$ it must be that player i defects. Note that $\Delta^C u_i > 0$ is equivalent to (5.6) and that $\Delta^C u_i < 0$ is equivalent to inequality (5.6) with rotated sign. □

Proof of Proposition 5.2. Consider a profile of moves $(a, p) \in J$ and a player i that contemplates whether she should deviate by playing $(a'_i, p'_i) \in J_i$. Let (a'_i, p'_i) strictly increase the direct value of player i,

$$\Delta^i \pi_i = \pi_i(a'_i, p'_i, a_{-i}, p_{-i}) - \pi_i(a, p) > 0 \; .$$

The corresponding change in utility of player i,

$$\Delta^i u_i = u_i^F(a_i', p_i', a_{-i}, p_{-i}) - u_i^F(a, p) ,$$

depends also on the effect her deviation has on the direct values of other players,

$$\Delta^i \pi_j = \pi_j(a_i', p_i', a_{-i}, p_{-i}) - \pi_j(a, p) .$$

a) Assume that by deviation of i the direct value of no other player increases by more than $\Delta^i \pi_i$,

$$\Delta^i \pi_j \leq \Delta^i \pi_i .$$

Then there are three possibilities for each player $j \neq i$: (i) $\pi_j(a,p) < \pi_i(a,p)$ and $\pi_j(a_i', p_i', a_{-i}, p_{-i}) < \pi_i(a_i', p_i', a_{-i}, p_{-i})$, (ii) $\pi_j(a,p) > \pi_i(a,p)$ and $\pi_j(a_i', p_i', a_{-i}, p_{-i}) \geq \pi_i(a_i', p_i', a_{-i}, p_{-i})$, or (iii) $\pi_j(a,p) \geq \pi_i(a,p)$ and $\pi_j(a_i', p_i', a_{-i}, p_{-i}) < \pi_i(a_i', p_i', a_{-i}, p_{-i})$. The difference in direct values between i and j changes when i deviates and this affects the change in the utility of player i: if (i) then the difference in direct values strictly increases by $\Delta^i \pi_i - \Delta^i \pi_j$, which decreases i's utility by $\frac{1}{n-1}\beta_i(\Delta^i \pi_i - \Delta^i \pi_j)$; if (ii) then the difference in direct values strictly decreases by $\Delta^i \pi_i - \Delta^i \pi_j$, which increases i's utility by $\frac{1}{n-1}\alpha_i(\Delta^i \pi_i - \Delta^i \pi_j)$; if (iii) then the difference may increase or decrease, by at most $\Delta^i \pi_i - \Delta^i \pi_j$, which changes i's utility between at least $-\frac{1}{n-1}\beta_i(\Delta^i \pi_i - \Delta^i \pi_j)$ and at most $\frac{1}{n-1}\alpha_i(\Delta^i \pi_i - \Delta^i \pi_j)$.

The utility of player i increases least if all her opponents are in (i) and most if all her opponents are in (ii). Hence,

$$\Delta^i \pi_i - \frac{1}{n-1}\beta_i \sum_{j \neq i}(\Delta^i \pi_i - \Delta^i \pi_j) \leq$$

$$\leq \Delta^i u_i \leq \Delta^i \pi_i + \frac{1}{n-1}\alpha_i \sum_{j \neq i}(\Delta^i \pi_i - \Delta^i \pi_j) . \quad \text{(A.4)}$$

b.) Take a player i that cooperates in (a^*, p^*), $a_i^* = C$. Let $\pi_i^{i:C}$ be her direct earning in (a^*, p^*). Say that i contemplates whether to defect and let (D, a_{-i}^*, p^*) be the profile of moves that coincides with (a^*, p^*) in all choices with exception of $a_i^* = D$. Let $\pi_i^{i:D}$ be the direct earning of i in (D, a_{-i}^*, p^*). Further, for each player j let $\pi_j^{i:C}$ and $\pi_j^{i:D}$ be their direct values in (a^*, p^*) and (D, a_{-i}^*, p^*). Let $g^* = g(p^*)$.

Let $\Delta^{i:D}\pi_j = \pi_j^{i:D} - \pi_j^{i:C}$ be the difference in direct values of player j if player i switches from cooperation to defection. By defection player i increases her own direct value by $\Delta^{i:D}\pi_i = 20 l_i(g^*)$, decreases the value of each of her neighbors $j \in L_i(g^*)$ by $\Delta^{i:D}\pi_j = -40$, and does not affect the direct values of other players.

Assume that i has at least one neighbor, $l_i(g^*) \geq 1$. Part a) of the proof now implies

$$\begin{aligned}
\Delta^{i:D}u_i &= u_i^F(D, a_{-i}^*, p^*) - u_i^F(a^*, p^*) \\
&\geq \Delta^{i:D}\pi_i - \frac{1}{n-1}\beta_i \sum_{j \neq i}(\Delta^{i:D}\pi_i - \Delta^{i:D}\pi_j) \\
&\geq 20l_i(g^*) - \frac{1}{n-1}\beta_i \left[\sum_{j \in L_i(g^*)} (20l_i(g^*) + 40) + \sum_{j \notin L_i(g^*) \cup \{i\}} 20l_i(g^*) \right] \\
&= 20l_i(g^*) - \frac{1}{n-1}\beta_i \left[(n-1)20l_i(g^*) - l_i(g^*)40\right] \\
&= 20l_i(g^*)(1 - \beta_i(1 - \frac{2}{n-1})).
\end{aligned}$$

Given that (a^*, p^*) is a Nash equilibrium in which player i cooperates it must be that $0 \geq \Delta u_i^{i:D}$, which implies $\beta_i \geq \frac{n-1}{n+1}$. □

Proposition A.1. *Consider utility function (5.11) and let γ_i satisfy (5.2). If $\tau_i^* = \max_{\tau_i \in (-\pi, \pi]} u_i(\tau_i)$ then $\tau_i^* \in (-\frac{\pi}{4}, \frac{\pi}{4})$. If $\gamma_i = 0$ then $\tau_i^* = 0$. If $\gamma_i = 1$ then $\tau_i^* = \frac{\pi}{4}$. If $-1 < \gamma_i < 1$ then $\gamma_i = \tan \tau_i$.*

Proof. An angle $\tau \in (\frac{\pi}{2}, \pi]$ (or $\tau \in [-\pi, -\frac{\pi}{2})$) should never be chosen because the angle $\pi - \tau$ (or $-\pi - \tau$) strictly increases utility (5.11): it maintains the earning of the other player and strictly increases own earning. To see this note that for any τ, $\sin(\pi - \tau) = \sin(-\pi - \tau) = \sin \tau$. Furthermore, if $\tau \in (\frac{\pi}{2}, \pi]$ then $(\pi - \tau) \in [0, \frac{\pi}{2})$, hence $\cos(\pi - \tau) > 0 > \cos \tau$, and if $\tau \in [-\pi, -\frac{\pi}{2})$ then $(-\pi - \tau) \in (-\frac{\pi}{2}, 0]$, hence $\cos(-\pi - \tau) > 0 > \cos \tau$. It is easy to see that any player strictly prefers angle 0 from angles $-\frac{\pi}{2}$ and $\frac{\pi}{2}$.

If angle τ_i^* maximizes the utility the first order condition must be satisfied:

$$\frac{du_i}{d\tau}(\tau_i^*) = -\sin \tau_i^* + \gamma_i \cos \tau_i^* = 0 \ .$$

This implies $\gamma_i = \tan \tau_i$. Given constraint (5.2), angles should thus be chosen from the interval $(-\frac{\pi}{4}, \frac{\pi}{4})$. □

Proposition A.2. *Consider utility function (5.13) and let α_i and β_i satisfy (5.7). If $\tau_i^* = \max_{\tau_i \in (-\pi, \pi]} u_i(\tau_i)$ then $\tau_i^* \in [0, \frac{\pi}{4}]$. If $\beta_i = 0$ then $\tau_i^* = 0$. If $\beta_i \geq 0.5$ then $\tau_i^* = \frac{\pi}{4}$. If $0 < \beta_i < 0.5$ then $\beta_i(\sin \tau_i + \cos \tau_i) = \sin \tau_i$.*

Proof. Let $\tau_i^* = \max_{\tau_i \in (-\pi,\pi]} u_i(\tau_i)$.

1.) Take an angle $\tau_i \notin [-\frac{\pi}{4}, \frac{3\pi}{4}]$, i.e., $\tau_i \in (-\frac{5\pi}{4}, -\frac{\pi}{4})$. We show that $u_i(-\frac{\pi}{2} - \tau_i) > u_i(\tau_i)$, for the angle $-\frac{\pi}{2} - \tau_i \in [-\frac{\pi}{4}, \frac{3\pi}{4}]$. Hence, no angle $\tau_i \notin [-\frac{\pi}{4}, \frac{3\pi}{4}]$ can maximize u_i and $\tau_i^* \in [-\frac{\pi}{4}, \frac{3\pi}{4}]$.

Elementary trigonometric equations imply

$$\cos(-\frac{\pi}{2} - \tau_i) = -\sin\tau_i \text{ and } \sin(-\frac{\pi}{2} - \tau_i) = -\cos\tau_i \ ,$$

hence

$$\sin(-\frac{\pi}{2} - \tau_i) - \cos(-\frac{\pi}{2} - \tau_i) = \sin\tau_i - \cos\tau_i$$

and

$$\cos(-\frac{\pi}{2} - \tau_i) - \cos\tau_i = -2\sin(\tau_i + \frac{\pi}{4}) > 0 \text{ for } \tau_i \in (-\frac{5\pi}{4}, -\frac{\pi}{4}) \ .$$

Therefore, for each angle $\tau_i \in (-\frac{5\pi}{4}, -\frac{\pi}{4})$ the angle $-\frac{\pi}{2} - \tau_i$ induces the same inequality between own and other's payoffs and strictly increases own payoff. Utility is thus strictly increased.

2.) Note next that $u_i(\frac{\pi}{4}) > u_i(\tau_i)$ for any angle $\tau_i \in (\frac{\pi}{4}, \frac{3\pi}{4}]$: at $\frac{\pi}{4}$ the inequality is minimized to 0, because $\cos\frac{\pi}{4} = \sin\frac{\pi}{4}$, and $\cos\tau_i < \cos\frac{\pi}{4}$ for $\tau_i \in (\frac{\pi}{4}, \frac{3\pi}{4}]$.

3.) Finally, we show that $u_i(-\tau_i) > u_i(\tau_i)$ for any $\tau_i \in [-\frac{\pi}{4}, 0)$, which together with 1.) and 2.) implies that $\tau_i^* \in [0, \frac{\pi}{4}]$. For $\tau_i \in [-\frac{\pi}{4}, 0)$:

$$\cos(\tau_i) = \cos(-\tau_i) > \sin(-\tau_i) = -\sin(\tau_i) > 0 \ ,$$

hence

$$\cos(\tau_i) - \sin(\tau_i) > \cos(\tau_i) + \sin(\tau_i) = \cos(-\tau_i) - \sin(-\tau_i) > 0 \ .$$

Again, the angle $-\tau_i$ induces the same own payoff as τ_i, but strictly decreases the loss of utility due to inequality. Utility is strictly higher with $-\tau_i$.

4.) Obviously, if $\beta_i = 0$, function $u_i(\tau_i) = \cos\tau_i$ is maximized with $\tau_i = 0$. For $\tau_i \in [0, \frac{\pi}{4})$, $\sin\tau_i + \cos\tau_i > 2\sin\tau_i$, hence

$$\frac{\mathrm{d}u_i}{\mathrm{d}\tau_i}(\tau_i) = -\sin\tau_i + \beta_i(\sin\tau_i + \cos\tau_i) > \sin\tau_i(2\beta_i - 1) \ .$$

If $\beta_i \geq 0.5$ then $\frac{du_i}{d\tau_i}(\tau_i) > 0$ for all $\tau_i \in [0, \frac{\pi}{4})$, and the continuous function $u_i(\tau_i)$ reaches its maximum over $[0, \frac{\pi}{4}]$ at $\tau_i = \frac{\pi}{4}$. If $0 < \beta_i < 0.5$ then $\mathrm{d}u_i/\mathrm{d}\tau_i(\tau_i) = 0$ implies $\beta_i(\sin\tau_i + \cos\tau_i) = \sin\tau_i$. □

A.3 Proofs from Chapter 6

Proof of Proposition 6.2. We begin by characterizing the payoff (6.1) of player i in terms of the distribution of types. Let $z(\mathbf{s}) = (n^{CA}(\mathbf{s}), n^{CE}(\mathbf{s}), n^{DA}(\mathbf{s}), n^{DE}(\mathbf{s}))$ be the distribution of types in state $\mathbf{s}$. For each j let T_j be the type of player j in state $\mathbf{s}$. Consider a player i that is updating her current strategy $s_i = s^{T_i}$ with a new strategy $s_i' = s^{T_i'}$ of type T_i'. For each player i, let $\delta_i^T(\mathbf{s}) = 1$ if her type in state $\mathbf{s}$ is T, and $\delta_i^T(s) = 0$ otherwise. The four-tuple $z(\mathbf{s}) - \delta_i(\mathbf{s}) = (n^{CA}(\mathbf{s}) - \delta_i^{CA}(\mathbf{s}), n^{CE}(\mathbf{s}) - \delta_i^{CE}(\mathbf{s}), n^{DA}(\mathbf{s}) - \delta_i^{DA}(\mathbf{s}), n^{DE}(\mathbf{s}) - \delta_i^{DE}(\mathbf{s}))$ gives the distribution of types among the players N/i in state $\mathbf{s}$.

The definition of the stage game payoff (6.2) may be rewritten as

$$\pi_i(a_i, p_i, a_{-i}, p_{-i}) = \sum_{j \in N/i} v(a_i, a_j) p_{ij} p_{ji} \tag{A.5}$$

for the case that the outside option value is $o = 0$, and as

$$\pi_i(a_i, p_i, a_{-i}, p_{-i}) = \sum_{j \in N/i} [v(a_i, a_j) p_{ij} p_{ji} + (1 - p_{ij} p_{ji}) o] \tag{A.6}$$

for general o. Let $p^{TT'}$ be 1 if players of type T propose links to players that were of type T' in the previous period, and 0 otherwise. Let a^T be the action chosen by strategy s^T. Combine the definition of payoff (6.1) with the definition of the stage game payoffs (A.5) to obtain

$$\begin{aligned} x_i(s_i' \mid \mathbf{s}) &= \sum_{j \in N/i} v(a_i(s_i'), a_j(s_j)) p_{ij}(s_i' \mid a(\mathbf{s})) p_{ji}(s_j \mid a(\mathbf{s})) \\ &\quad + (r-1) \sum_{j \in N/i} v(a_i(s_i'), a_j(s_j)) p_{ij}(s_i' \mid a(\mathbf{s}')) p_{ji}(s_j \mid a(\mathbf{s}')) \\ &= \sum_{T \in \mathcal{T}} \left(n^T(\mathbf{s}) - \delta_i^T(\mathbf{s})\right) v(a^{T_i'}, a^T) p^{T_i' T} [p^{T T_i} + (r-1) p^{T T_i'}] \end{aligned} \tag{A.7}$$

for the case $o = 0$. One can show in the similar way that for general o,

$$\begin{aligned} x_i(s_i' \mid \mathbf{s}) = \sum_{T \in \mathcal{T}} \left(n^T(\mathbf{s}) - \delta_i^T(\mathbf{s})\right) \Big[& v(a^{T_i'}, a^T) p^{T_i' T} [p^{T T_i} + (r-1) p^{T T_i'}] + \\ & + o\left(1 - p^{T_i' T} [p^{T T_i} + (r-1) p^{T T_i'}]\right) \Big] \,. \end{aligned} \tag{A.8}$$

Let

$$\widehat{v}[T_i, T_i' \mid T] = v(a^{T_i'}, a^T) p^{T_i'T} [p^{TT_i} + (r-1) p^{TT_i'}] \\ + o\left(1 - p^{T_i'T} [p^{TT_i} + (r-1) p^{TT_i'}]\right) \quad \text{(A.9)}$$

be the r-period payoff to any player i that changed her type from T_i to T_i' from the interaction with any player $j \neq i$ of type T. This payoff is independent of the identities i and j. Equation (A.8) rewrites as

$$x_i(s_i' \mid \mathbf{s}) = \sum_{T \in \mathcal{T}} \left(n^T(\mathbf{s}) - \delta_i^T(\mathbf{s})\right) \widehat{v}[T_i, T_i' \mid T] \; .$$

The best response of each player thus depends on her own previous type and on the distribution of types among the remaining players. Consequently, players who were playing the same strategy in the previous period have the same set of best responses, which depends only on the distribution of the types in the previous state. If state $\mathbf{s} = (s_1, ..., s_n)$ is absorbing, than the state $\mathbf{s}^\phi = (s_{\phi(1)}, ..., s_{\phi(n)})$, for any permutation of the player indices $\phi : N \to N$, is also absorbing. To conclude the proof, note that S_z consists exactly of all states $\mathbf{s}^\phi$ that can be obtained by permuting the player indices via any permutation ϕ, from any state $\mathbf{s} \in S_z$. □

Proof of Proposition 6.5. Several observations can be made on the basis of (A.7). First, non-exclusive strategy always earns at least as much as its exclusive counterpart: for all $\mathbf{s}$,

$$x_i(s_i^{CA} \mid \mathbf{s}) \geq x_i(s_i^{CE} \mid \mathbf{s}) \text{ and } x_i(s_i^{DA} \mid \mathbf{s}) \geq x_i(s_i^{DE} \mid \mathbf{s}) \; . \quad \text{(A.10)}$$

This is because $p^{CA,T} \geq p^{CE,T}$ and $p^{DA,T} \geq p^{DE,T}$ for all T, with strict inequalities for $T \in \{DA, DE\}$. Equality is attained only in the border cases: $x_i(s_i^{CA} \mid \mathbf{s}) = x_i(s_i^{CE} \mid \mathbf{s})$ implies that $n^{DA} = n^{DE} = 0$. Similarly, $x_i(s_i^{DA} \mid \mathbf{s}) = x_i(s_i^{DE} \mid \mathbf{s})$ implies $n^{DA} = 0$.

1.] Strategy $\mathbf{s}^{DE}$ may thus be a part of an absorbing state only when $n^{DA} = n^{CE} = 0$. Consider a state with type distribution $(n-p, 0, 0, p)$. If $r = 1$ cooperation is not a best response, implying $p = n$. For $r \geq 2$ exclusion is not best response in presence of another defector, implying $p \leq 1$. For $p = 1$ cooperation is a best response if and only if condition (6.7) holds.

2.] When there are no exclusive players, that is when $n^{CE} = n^{DE} = 0$, the utility (A.7) rewrites as

$$x_i(s_i' \mid \mathbf{s}) = \sum_{T \in \{CA, DA\}} n^T v(a^{T_i'}, a^T) p^{T_i'T} r$$

in which case the strategy s^{DA} is the best response. To see this note that $v(D, a^T) > v(C, a^T)$ for any T, which in case that in $\mathbf{s}$ there are no exclusive players implies $x_i(s_i^{DA} \mid \mathbf{s}) \geq x_i(s_i^{CA} \mid \mathbf{s})$. Combine this with (A.7) to conclude that s^{DA} maximizes the utility.

3.] When $n^{CE} \geq 1$, there are no defective players, $n^{DA} = n^{DE} = 0$. A state with type distribution $(p, n-p, 0, 0)$ is absorbing whenever players playing strategy s^{CA} do not receive higher utility (A.7) by updating to strategy s^{DA}. A straightforward calculation yields that for a player i with $s_i = s^{CA}$, $x_i(s_i^{CA} \mid \mathbf{s}) \geq x_i(s_i^{DA} \mid \mathbf{s})$ if and only if condition (6.8) holds.

Absorbing states may therefore have only one of the following type distributions: $(0, 0, 0, n)$ only for $r = 1$, $(n-1, 0, 0, 1)$ whenever (6.7), $(0, 0, n, 0)$, and $(p, n-p, 0, 0)$ whenever (6.8). Constraint $f > c > d > e$ implies that conditions (6.7) and (6.8) do not hold for $r = 1$.

Finally, note that for $p = 0$ condition (6.8) rewrites to (6.6). □

Proof of Propositions 6.6 and 6.7. Let $\widehat{v}[T_i, T_i' \mid T]$ be defined as in (A.9) and let

$$\overline{v}[T \mid T_i, T_i'] = v(a^T, a^{T_i'}) p^{T_i' T} [p^{T T_i} + (r-1) p^{T T_i'}] \\ + o\left(1 - p^{T_i' T} [p^{T T_i} + (r-1) p^{T T_i'}]\right)$$

be the r-period payoff to any player j of type T from interaction with player $i \neq j$ that changed her type from T_i to T_i'. When $o = 0$, the r-period utility of player j, $x_{ij}(s_i' \mid \mathbf{s})$, defined as in (6.11), can be rewritten as

$$\begin{aligned} x_{ij}(s_i' \mid \mathbf{s}) &= r \sum_{k \in N/\{i,j\}} v(a_j(s_j), a_k(s_k)) p_{jk}(s_j \mid a(\mathbf{s})) p_{kj}(s_k \mid a(\mathbf{s})) \\ &\quad + v(a_j(s_j), a_i(s_i')) p_{ji}(s_j \mid a(\mathbf{s})) p_{ij}(s_i' \mid a(\mathbf{s})) \\ &\quad + (r-1) v(a_j(s_j), a_i(s_i')) p_{ji}(s_j \mid a(\mathbf{s}')) p_{ij}(s_i' \mid a(\mathbf{s}')) \\ &= \sum_{T \in \mathcal{T}} \left(n^T(\mathbf{s}) - \delta_j^T(\mathbf{s}) - \delta_i^T(\mathbf{s})\right) r v(a^{T_j}, a^T) p^{T_j T} p^{T T_j} \\ &\quad + v(a^{T_j}, a^{T_i'}) p^{T_i' T_j} [p^{T_j T_i} + (r-1) p^{T_j T_i'}] \end{aligned}$$

again using the definition of the stage game payoffs (A.5), which further implies

$$x_{ij}(s_i' \mid \mathbf{s}) = \sum_{T \in \mathcal{T}} \left(n^T(\mathbf{s}) - \delta_j^T(\mathbf{s}) - \delta_i^T(\mathbf{s})\right) \overline{v}[T_j \mid T, T] + \overline{v}[T_j \mid T_i, T_i']$$

and

$$
\begin{aligned}
x_i(s_i' \mid \mathbf{s}) - x_{ij}(s_i' \mid \mathbf{s}) = &\sum_{T \in \mathcal{T}} \left(n^T(\mathbf{s}) - \delta_i^T(\mathbf{s})\right) \widehat{v}[T_i, T_i' \mid T] \\
&- \sum_{T \in \mathcal{T}} \left(n^T(\mathbf{s}) - \delta_i^T(\mathbf{s})\right) \overline{v}[T_j \mid T, T] \\
&+ \overline{v}[T_j \mid T_j, T_j] - \overline{v}[T_j \mid T_i, T_i'] \ .
\end{aligned}
$$

One can show in the similar fashion that the above equation holds for general o. It is important to note that players, distinct from i, of the same type have the same utility: if $T_j = T_k$ for some $j, k \neq i$, then $x_{ij}(s_i' \mid \mathbf{s}) = x_{ik}(s_i' \mid \mathbf{s})$, and

$$
x_i(s_i' \mid \mathbf{s}) - x_{ij}(s_i' \mid \mathbf{s}) = x_i(s_i' \mid \mathbf{s}) - x_{ij}(s_i' \mid \mathbf{s}) \ . \qquad \text{(A.11)}
$$

This motivates the use of notation

$$
x_{iT}(s_i' \mid \mathbf{s}) = \sum_{\overline{T} \in \mathcal{T}} \left(n^{\overline{T}}(\mathbf{s}) - \delta_j^{\overline{T}}(\mathbf{s}) - \delta_i^{\overline{T}}(\mathbf{s})\right) \overline{v}[T \mid \overline{T}, \overline{T}] + \overline{v}[T \mid T_i, T_i']
$$

for each type T. Clearly, $x_{ij}(s_i' \mid \mathbf{s}) = x_{iT_j}(s_i' \mid \mathbf{s})$.

1.] Assume the model of altruism of Sec 6.5.1 and let $\gamma_i = \gamma$ for each player i. In this case the definition of utility (6.12) reduces to

$$
\begin{aligned}
u_i^A(s_i' \mid \mathbf{s}) &= x_i(s_i' \mid \mathbf{s}) + \gamma \sum_{j \neq i} x_{ij}(s_i' \mid \mathbf{s}) \\
&= x_i(s_i' \mid \mathbf{s}) + \gamma \sum_{T \in \mathcal{T}} \left(n^T(\mathbf{s}) - \delta_i^T(\mathbf{s})\right) x_{iT}(s_i' \mid \mathbf{s}) \ .
\end{aligned}
$$

The value of $u_i^A(s_i' \mid \mathbf{s})$ therefore depends on the previous and the updated strategies of player i, and on the distribution of types among the remaining players. In particular, it does not depend on the identities of players of particular type. This means that if two players were of the same type in the previous period, the sets of their best responses coincide. Following the final steps outlined in the proof of Prop. 6.2 we prove the claim of Prop. 6.6.

2.] Assume the model of inequity aversion of Sec 6.5.2 and let $\alpha_i = \alpha$ and $\beta_i = \beta$ for each player i. The definition of utility (6.15) reduces to

$$
\begin{aligned}
u_i^F(s_i' \mid \mathbf{s}) = x_i(s_i' \mid \mathbf{s}) &\\
&- \frac{\alpha}{n-1} \sum_{T \in \mathcal{T}} \left(n^T(\mathbf{s}) - \delta_i^T(\mathbf{s})\right) \max\{x_{iT}(s_i' \mid \mathbf{s}) - x_i(s_i' \mid \mathbf{s}), 0\} \\
&- \frac{\beta}{n-1} \sum_{T \in \mathcal{T}} \left(n^T(\mathbf{s}) - \delta_i^T(\mathbf{s})\right) \max\{x_i(s_i' \mid \mathbf{s}) - x_{iT}(s_i' \mid \mathbf{s}), 0\}.
\end{aligned}
$$

Again, the value of $u_i^F(s_i' \mid \mathbf{s})$ depends on the previous and the updated strategies of player i, and on the distribution of types among the remaining players. In particular, it does not depend on the identities of players of particular type and the claim of Prop. 6.7 follows in the similar steps as above. □

B

Appendix: Experimental Instructions

This appendix contains the precise text of the instructions that were given to the subjects. Our comments to the instructions are written in italicized text and between square brackets, as in '[*comments*]', and describe the differences between the instructions across our three treatments.

Introduction

Welcome to this experiment on decision-making. In this experiment you can earn money. How much you earn depends on your decisions and the decisions of other participants.

The experiment consists of two completely independent parts. In each part you can earn points. Your earnings in each part are independent of your earnings in the other part. At the end of the experiment you get paid your earned points privately in cash, according to the exchange rate:

5 points = 1 eurocent.

[*In treatment II 5 points were worth 2 eurocents.*]

Now you will receive the instructions for the first part. You will get the instructions for the second part only after the first part is finished. At the end of the experiment you will be asked to fill in a short questionnaire. Thereafter you will be paid your earnings.

During the whole experiment, you are not allowed to communicate with other participants in whatever way.

If you have a question, please raise your hand. We will then come to you to answer it.

Instructions part I

In part I of the experiment you are asked to make one decision. This decision is about assigning an amount (of points) to yourself and an amount to another, randomly chosen participant.

You received a print-out of the computer screen for part I from us. Please take it in front of you now. We shall explain your choice options with the help of this print-out.

On the screen you see a **circle**. By choosing a point on this circle you decide about the assignment of an amount to yourself and of another amount to one other participant in this experiment. This other participant will be chosen completely randomly by the computer.

Each point on the circle represents an amount that is added to your earnings (+) or is subtracted from your earnings (-) and an amount that is added to the earnings of the other participant (+) or is subtracted from his or her earnings (-). Hence, with your decision you can increase or decrease your earnings and the earnings of the other participant.

You can click on any point of the circle with the help of your mouse. You will then see an arrow that shows the decision you made. Additionally, you will see the corresponding amounts (points) in the window 'decision', at the right of the circle. In the **For me** field you will see the amount that you assigned to yourself, and in the **For the other** field the amount you assigned to the other participant. To change your decision you can click on another point of the circle. You may also use the buttons below the circle. With these buttons you can refine your decision: you can move the arrow a little bit in the clockwise direction (left button) or in the counter-clockwise direction (right button). If you are satisfied with your decision you have to confirm it by clicking on the button '**Confirm**'. A '**last chance**' window will then appear. Click on '**yes**' if you do not want to change your choice anymore.

Your earnings will depend on your own decision and the decision of the other, randomly chosen, participant you are paired with. This other participant has to make a similar decision. Your earnings in this part of the experiment are therefore equal to the sum of the amount you assign to yourself and the amount that the other participant assigns to you. Note: these amounts can be positive but also negative. In the latter case the amount will be subtracted from the earnings of the respective recipient.

During the whole experiment, you will receive no information about the decision of the other participant you are paired with. Only after the end of the experiment you will get informed about the amount this other participant has assigned

to you. The other participant will also get informed about the amount you assign to her or him only after the end of the experiment.

If you have any questions now, please raise your hand. If you do not have any questions, please click on '**READY**'. Part I of the experiment will begin shortly. **There is no practice round**.

[*Part I started after all subjects pressed 'READY'. Instructions for part II were given only after all subjects completed part I.*]

Instructions part II

In this part of the experiment every participant is in a **group of six** with five other participants. Before part II begins, these groups of six will be formed arbitrarily with the help of the computer. The group you are in will not change during the experiment. You will get no information about the identity of the persons in your group, nor during the experiment, neither after the experiment. Neither will the other participants receive any information about your identity.

Each person in your group is indicated by a letter. The other five persons in your group will be indicated by the letters **A**, **B**, **C**, **D** and **E**. You will receive the name **Me**. The same letter always refers to the same person.

This part of the experiment consists of 60 rounds. In each round you can earn points. Your total earnings in this part of the experiment is the sum of your earnings in each of the 60 rounds.

In each round, you – and each other person in your group – has to make two decisions. You have to make a **decision: connections** and a **decision: color**, explained in detail below. [*In treatment III the following sentence replaced the above paragraph:*] In each round, you – and each other person in your group – has to make one decision, explained in detail below.

Note: during all 60 rounds the other participants in your group will stay the same persons.

Decisions (in one single round)

decision: connections

You have to decide with whom you want to make a connection. You can choose any other person in your group and you can make as many connections as you want. These choices – together with choices of the other persons in your group – determine with whom you **interact** (in the respective round) as explained below:

- You will interact with a person in your group only if you make a connection with him or her **and** this person also makes a connection with you.
- You will not interact with persons with whom you do not make connections.
- You will not interact with persons who do not make connections with you.

For your convenience we will call those persons in your group with whom you interact: **your neighbors**. Your neighbors are therefore those persons with whom you made a connection **and** who made a connection with you.

[*In treatment II the above subsection "**decision: connections**" was replaced by the following sentence:*] For your convenience we will call the five other persons in your group: **your neighbors**.

decision: color

Each person in your group has to choose between two colors: **blue** and **green**. As explained above, you interact with each of your neighbors. In these interactions you can earn points. The color chosen by you and the color chosen by your neighbors determine how much you and your neighbors earn in these interactions.

Your earnings and the earnings of a neighbor of you are determined as follows:

your color	color of your neighbor	your earning	earning of your neighbor
blue	blue	50 points	50 points
blue	green	10 points	70 points
green	blue	70 points	10 points
green	green	30 points	30 points

In each round you will interact with all your neighbors. Therefore, your total earnings in a round from interactions with your neighbors is the sum of all points you earn in each of the interactions. [*The following two sentences are given only in treatments I and II:*] For each person in your group with whom you do not interact (i.e. all persons who are not your neighbors) you earn 0 points. For example, if you have no neighbors in a round, then you earn 0 points in this round.

Note: you can not choose different colors for different neighbors. You can change, however, the color as well as your connections between the rounds.[*In treatment III this sentence was replaced by the following sentence:*]You can change, however, the color between the rounds.

You get now information about the computer screen that you will see soon. You received a print-out of the computer screen (**Example screen 1**) from us. Take this print-out in front of you. You see now also the color version of it as an overhead projection. The screen consists of five windows: **round**, **decision: connections** [*This window is mentioned only in treatments I and II.*], **decision: color**, **information** and **special explanation**.

- **round**: This window holds information about past round(s). At the beginning of a new round you will receive information about decisions in the previous round, automatically there (In the example, this is round 2; see upper left.). In the window there are 6 little squares, named **Me**, **A**, **B**, **C**, **D** and **E**. **Me** always refers to you. The letters refer to the other five persons in your group.
  - [*This item appears only in treatments I and II.*] A thick black line between two persons (letters or '**Me**') indicate that these two persons were neighbors in the previous round, that is, they had interaction with each other (see the line between **Me** and **E** on the example screen).
  - [*This item appears only in treatments I and II.*] A thin gray line between two persons indicates that only one of them made a connection: such a gray line is full on the side where the connection was made, and broken on the side where connection was not made. (See, e.g., the line between **Me** and **B** on the example screen: **Me** made a connection with **B**, but **B** did not make a connection with **Me**.)
  - The squares are either **blue** or **green**. This gives you information about the color choices of the persons in your group. (In the example: persons **B**, **C** and **E** have chosen the color **green**, while the persons **Me**, **A** and **D** have chosen the color **blue**.)
  - At the bottom of this window you find two buttons called **previous** and **next**. You can use these buttons to look at the decisions in all previous rounds. Between the buttons you can find your **earnings** (in points) in the corresponding round.
- **decision: connections** [*This item appears only in treatments I and II.*] In this window you can choose with whom you want to make a connection. Again, there are five squares named **A**, **B**, **C**, **D** and **E** and the square **Me**. You can make a connection with another person in your group by clicking on the corresponding square. A thin gray line will appear to indicate that you made a connection. To

remove a connection you made, click on the corresponding square again. (On the example screen: **Me** made a connection with persons **B** and **E**.)
Note: At the beginning of each round you will always see the connections that you made in the previous round. If you want, you can remove these connections in the way described above.

- **decision: color** This window is located precisely underneath the window just described. In this window you choose between the two colors **blue** and **green**. You make your decision by clicking in the empty button to the left of **blue** or **green**. When you are satisfied with all your decisions (that is, with **both**: the connections you made and the chosen color), you have to confirm these decisions by clicking on the button '**Confirm**'. You will then see a '**last chance**' screen.
- **information** In this window you find information about the current round and about your **total earnings** up to this round.
- **special explanation** If this window becomes yellow you have to make your decisions. After you have made and confirmed your decisions this window will turn gray. You then have to wait until all participants are ready. Only then a new round will begin.

Information: After the round is over you will receive information about all the choices (that is, all connections made and all color choices) of all persons in your group. Similarly, all other persons in your group will receive information about all your decisions.

This is the end of the instructions. When all participants are ready you will have to answer a few questions, to make sure that you understood the instructions properly. Thereafter, there will be one **practice round** in which you cannot yet earn any money. Your decisions in this practice round will not be revealed to other participants. Only after the practice round is over part II of the experiment will begin.

If you have any questions please raise your hand. If you have no questions, press '**ready**'.

[*After all subjects clicked on 'READY' each had to answer, individually at her computer terminal, a set of questions designed to test her comprehension of the instructions. The practice round began only after each subject correctly answered all the questions. Part II of the experiment started after all subjects finished with the practice round.*]

References

[1] Dawes R.M. (1980) Annual Review of Psychology **31**:169–193
[2] Axelrod R. (1984) *The Evolution of Cooperation* New York: Basic Books
[3] Hardin G. (1968) Science **162**:1243–1248
[4] Stiglitz J.E. (1999) in: *Global Public Goods* Oxford University Press, New York 326–343
[5] Hobbes T. (1651) *The Leviathan* Penguin Classics; New edition (1982)
[6] Mauss M. (1967) *The gift* Norton, New York
[7] Sahlins M. (1972) *Stone age economics* Aldine-Atherton, Chicago
[8] Fafchamps M., Lund S. (2003) Journal of Development Economics **71**:261–287 sSRN working paper 97-014
[9] Healey C. (1990) *Maring hunters and traders: production and exchange in the Papua New Guinea highlands* University of California Press, Berkeley
[10] Lomnitz L.A. (1977) *Networks and marginality: Life in a Mexican shantytown* Academic Press, New York
[11] Coleman J.S. (1988) American Journal of Sociology **94**:95–120
[12] Putnam R.D. (2000) *Bowling Alone: The Collapse and Revival of American Community* Simon & Schuster, New York
[13] Friedman J.W. (1971) Review of Economic Studies **38**:1–12
[14] Kreps D.M., Milgrom P., Roberts J., Wilson R. (1982) Journal of Economic Theory **27**:245–252
[15] Coleman J.S. (1984) AEA Papers and Proceedings :84–88
[16] Granovetter M. (1985) American Journal of Sociology **91**:481–510
[17] Nowak M.A., May R.M. (1992) Nature **359**:826–829

[18] Eshel I., Samuelson L., Shaked A. (1998) American Economic Review **88**:157–179

[19] Cooper R., DeJong D.V., Forsythe R. (1996) Games and Economic Behavior **12**:187–218

[20] Huck S., Normann H.T., Oechssler J. (2004) Journal of Economic Behavior and Organization **53**:435–446

[21] Cassar A. (2002) Coordination and cooperation in local, random and small world networks:experimentalevidence proceedings of the 2002 North American Summer Meetings of the Econometric-Society:Game Theory.

[22] Kirchkamp O., Nagel R. (2000) Local and group interaction in prisoners' dilemma experiments - imitatelocally,think globally sFB 504 discussion paper No. 00-11

[23] Orbell J.M., Schwartz-Shea P., Simmons R.T. (1984) The American Political Science Review **78**:147–162

[24] Hauk E., Nagel R. (2001) Journal of conflict resolution **45 (6)**:770–793

[25] Orbell J.M., Dawes R.M. (1993) American Sociological Review **58**:787–800

[26] von Neumann J., Morgenstern O. (1944) *Theory of Games and Economic Behavior* Princeton University Press, Princeton

[27] Myerson R.B. (1991) *Game Theory: Analysis of Conflict* Harvard University Press, Cambridge, MA

[28] Nash J. (1951) Annals of Mathematics **54**:289–295

[29] van Damme E. (1987) *Stability and Perfection of Nash Equilibria* Springer Verlag, Berlin

[30] Fudenberg D., Tirole J. (1991) *Game Theory* MIT Press, Cambridge, Massachusetts

[31] Selten R. (1965) Zeitschrift für die gesamte Staatswissenschaft **12**:301–324

[32] Osborne M.J., Rubinstein A. (1994) *A Course in Game Theory* MIT Press, Cambridge, Massachusetts

[33] Benoit J.P., Krishna V. (1985) Econometrica **53**:905–922

[34] Friedman J.W. (1985) Journal of Economic Theory **35**:390–398

[35] Palfrey T.R., Rosenthal H. (1984) Journal of Public Economics **24**:171–193

[36] Berge C. (1973) *Graphs and Hypergraphs* North-Holland

[37] Freeman L. (1979) Social Networks **1**:215–239

[38] Wasserman S., Faust K. (1994) *Social Network Analysis: Methods and Applications* Cambridge University Press

[39] Tijms H.C. (1994) *Stochastic Models: An Algorithmic Approach* Wiley

[40] Bhattacharya R.N., Waymire E.C. (1990) *Stochastic Processes with Applications* John Wiley & Sons, New York

[41] Jackson M.O. (2004) in: G. Demange, M. Wooders (eds.) *Group Formation in Economics: Networks, Clubs and Coalitions* Cambridge University Press, Cambridge

[42] Ule A. (2005) *Exclusion and Cooperation in Networks* Tinbergen Institute Research Series Tinbergen Institute, Amsterdam

[43] Droste E., Gilles R.P., Johnson C. (2000) Evolution of conventions in endogenous social networks mimeo: Virginia Tech

[44] Goyal S., Vega-Redondo F. (2005) Games and Economic Behavior **50**:178–207 mimeo: Erasmus University

[45] Jackson M.O., Watts A. (2002) Games and Economic Behavior **41**:265–291

[46] Corbae D., Duffy J. (2003) Experiments with network economies mimeo

[47] Hauk E. (2001) Computational Economics **18**:65–87

[48] Smucker M.D., Ann Stanley E., Ashlock D. (1994) Analyzing social network structures in the iterated prisoner's dilemma withchoiceand refusal Tech. Rep. CS-TR-94-1259 University of Wisconsin-Madison, Department of Computer Sciences

[49] Hanaki N., Peterhansl A., Dodds P.S., Watts D. (2004) Cooperation in evolving social networks mimeo

[50] Lyon F. (2000) World Development **28**:663–681

[51] Huck S., Tyran J.R. (2004) Reciprocity, social ties, and competition in markets for experience goods Institute of Economics at University of Copenhagen working paper 04-12

[52] Ranganathan K., Ripeanu M., Sarin A., Foster I. (2003) to share or not to share - an analysis of incentives to contribute incollaborativefile sharing environments presented at the first Workshop on the Economics of Peer-to-Peer Systems,Berkeley

[53] Morselli R., Katz J., Bhattacharjee B. (2004) A game-theoretic framework for analyzing trust-inference protocols presented at the second Workshop on Economics of Peer-to-Peer Systems, atHarvard

[54] Kung H.T., Wu C.H. (2003) Differentiated admission for peer-to-peer systems: incentivizing peers tocontributetheir resources presented at the first Workshop on the Economics of Peer-to-Peer Systems,atBerkeley

[55] Galeotti A., Goyal S., Kamphorst J. (2005) Games and Economic Behavior Forthcoming
[56] Goeree J.K., Riedl A., Ule A. (2003) In search of stars: An experimental study of endogenous network formation cREED mimeo
[57] Haller H., Sarangi S. (2005) Mathematical Social Sciences **50**:181–201
[58] Bala V., Goyal S. (2000) Econometrica **68**:1131–1230
[59] Jackson M.O., Wolinsky A. (1996) Journal of Economic Theory **71**:44–74
[60] Watts A. (2001) Games and Economic Behavior **34**:331–341
[61] Jackson M.O., Watts A. (1999) The evolution of social and economic networks Tech. Rep. wp 1044 CALTECH
[62] Gilles R.P., Sarangi S. (2004) Network formation under consent mimeo
[63] Aumann R., Myerson R.B. (1988) in: *The Shapley value* Cambridge University Press, Cambridge, United Kingdom 175–191
[64] Curarini S., Morelli M. (2000) Review of Economic Design **5**:229–249
[65] Deroïan F. (2003) Economics Letters **80**:343–349
[66] Dutta B., Ghosal S., Ray D. (2005) Journal of Economic Theory **122**:143–164
[67] Watts A. (2002) Economics Letters **74**:277–282
[68] Auman R., Dréze J. (1974) International Journal of Game Theory **3**:217–237
[69] Myerson R.B. (1977) Mathematics of Operations Research **2**:225–229
[70] van den Nouweland A. (2004) The formation of communication networks in cooperative games mimeo
[71] Slikker M., van den Nouweland A. (2000) Review of Economic Design **5**:333–362
[72] Skyrms B., Pemantle R. (2000) Proceedings of the National Academy of Sciences **97**:9340–9346
[73] Aumann R.J., Shapley L.S. (1994) in: N. Megiddo (ed.) *Essays in Game Theory: In Honor of Michael Maschler* Springer-Verlag 1–15 (originaly mimeo 1976)
[74] Fudenberg D., Maskin E. (1986) Econometrica **54**:533–554
[75] Neyman A. (1985) Economics Letters **19**:227–229
[76] Calvó-Armengol A. (2004) Journal of Economic Theory **115**:191–206
[77] Goyal S., Joshi S. (2003) Unequal connections mimeo

[78] Goyal S., Vega-Redondo F. (2004) Structural holes in social networks mimeo

[79] Smith L. (1995) Econometrica **63**:425–430

[80] Benoit J.P., Krishna V. (2000) The folk theorems for repeated games: A synthesis mimeo

[81] Galeotti A., Meléndez-Jiménez M.A. (2004) Exploitation and cooperation in networks tinbergen Institute Discussion Paper 04-076/1

[82] Hirshleifer D., Rasmusen E. (1989) Journal of Economic Behavior and Organization **12(1)**:87–106

[83] Dutta B., Van der Nouweland A., Tijs S. (1998) International Journal of Game Theory **27**:245–256

[84] Dutta B., Mutuswami S. (1997) Journal of Economic Theory **76**:322–344

[85] Bernheim B., Peleg B., Whinston M. (1987) Journal of Economic Theory **42**:1–12

[86] Aumann R. (1959) in: A. Tucker, R. Luce (eds.) *Contributions to the Theory of Games IV* Princeton University Press 287–324

[87] Jackson M.O., van der Nouweland A. (2002) Strongly stable networks mimeo

[88] Slikker M., van den Nouweland A. (2001) Games and Economic Behavior **34**:153–175

[89] Kleinberg J.M. (2000) Nature **406**:845

[90] Milgram S. (1976) Psychology Today **2**:60–67

[91] Newman M.E.J., Barabási A.L., Watts D.J. (2003) *The Structure and Dynamics of Networks* Princeton University Press, Princeton

[92] Watts D.J. (2003) *Six Degrees: The Science of Connected Age* Norton, New York

[93] Anderson S.P., Goeree J.K., Holt C.A. (1998) Journal of Public Economics **70**:297–323

[94] Axelrod R. (1987) in: L. Davis (ed.) *Genetic Algorithms and Simulated Annealing* London: Pittman 32–41

[95] Fehr E., Schmidt K.M. (1999) The Quarterly Journal of Economics **114**:817–868

[96] Fehr E., Gächter S. (2000) The American Economic Review **90**:980–994

[97] Berg J., Dickhaut J., McCabe K. (1995) Games and Economic Behavior **10**:122–142

[98] Brandts J., Schram A. (2000) Journal of Public Economics **8-9**:399–427

[99] Fehr E., Kirchsteiger G., Riedl A. (1993) Quarterly Journal of Economics **108**:437–460
[100] Fischbacher U., Gächter S., Fehr E. (2001) Economics Letters **71**:397–404
[101] Güth W., Schmittberger R., Schwarze B. (1982) Journal of Economic Behavior and Organization **3**:367–388
[102] Seinen I., Schram A. (2001) Social status and group norms: Indirect reciprocity in a helping experiment forthcoming in *European Economic Review*
[103] Fehr E., Gächter S. (1998) European Economic Review **42**:845–859
[104] Roth A.E., Prasnikar V., Okuno-Fujiwara M., ShmuelZamir (1991) American Economic Review **81**:1068–1095
[105] Egas M., Riedl A. (2004) The economics of altruistic punishment and the demise of cooperation cREED mimeo
[106] Berninghaus S., Ehrhart K.M., Ott M. (2006) Experimental Economics **9**:237–251
[107] Callander S., Plott C. (2005) Journal of Public Economics **89**:1469–1495
[108] Kosfeld M. (2004) Review of Network Economics **3**:20–42
[109] Cook K.S., Emerson R.M. (1978) American Sociological Review **43**:721–739
[110] Cook K.S., Emerson R.M., Gilmore M.R., Yamagishi T. (1983) American Journal of Sociology **89**:275–305
[111] Lawler E.J., Yoon J. (1996) American Sociological Review **61**:89–108
[112] Hauk E. (1999) Multiple prisoner's dilemma games with(out) an outside option: An experimentalstudy mimeo
[113] Coricelli G., Fehr D., Fellner G. (2004) Journal of Conflict Resolution **48**:356–378
[114] Brown M., Falk A., Fehr E. (2004) Econometrica **72**:747–780
[115] Page T., Putterman L., Unel B. (2002) Voluntary association in public goods experiments: Reciprocity, mimicry,andefficiency mimeo
[116] Gunnthorsdottir A., Houser D., McCabe K., Ameden H. (2000) Excluding free-riders improves reciprocity and promotes the private provisionofpublic goods mimeo
[117] Swope K.J. (2002) Experimental Economics **5**:209–222
[118] Ehrhart K.M., Keser C. (1999) Mobility and cooperation: On the run working Paper 99s-24, CIRANO, Montreal

[119] Goeree J.K., Holt C.A. (2001) American Economic Review **91**:1402–1422
[120] Kagel J.H., Roth A.E. (eds.) (1995) *The Handbook of Experimental Economics* Princeton University Press
[121] Bergstrom T.C. (2002) Journal of Economic Perspectives **16**:67–88
[122] Andreoni J. (1988) Journal of Public Economics **37**:291–304
[123] Isaac R.M., Walker J., Thomas S. (1984) Public Choice **43**:113–149
[124] Isaac R.M., Walker J.M., Williams A.W. (1994) Journal of Public Economics :1–36
[125] Brandts J., Riedl A., van Winden F. (2004) Competition and well-being mimeo
[126] Weibull J.W. (2000) Testing game theory stockholm School of Economics working paper 382
[127] Bolton G., Ockenfels A. (2000) American Economic Review **90**:166–193
[128] Cox J.C., Friedman D., Gjerstad S. (2004) A tractable model of reiprocity and fairness mimeo
[129] Dufwenberg M., Kirchsteiger G. (1999) A theory of sequential reciprocity mimeo, CentER for Economic Research, Tilburg
[130] Falk A., Fischbacher U. (1999) A theory of reciprocity working paper No. 6, University of Zürich
[131] Rabin M. (1993) American Economic Review **83**:1281–1302
[132] Levine D.K. (1998) Review of Economic Dynamics **1**:593–622
[133] Ledyard J.O. (1995) in: J.H. Kagel, A.E. Roth (eds.) *Handbook of Experimental Economics* Princeton: Princeton University Press 111–194
[134] Isaac R.M., Walker J. (1988) Quarterly journal of economics **103**:179–200
[135] Eckel C.C., Grossman P.J. (1996) Games and Economic Behavior **16**:181–191
[136] Forsythe R., Kennan J., Sopher B. (1991) American Economic Review **81**:253–278
[137] Sonnemans J., van Dijk F., van Winden F. (2004) On the dynamics of social ties structures in groups mimeo
[138] Griesinger D.W., Livingston J.W. (1973) Beavioral Science **18**:173–188
[139] Liebrand W.B.G. (1984) European Journal of Social Psychology **14**:239–264

[140] Offerman T. (1997) *Beliefs and Decision Rules in Public Good Games* Kluwer, Dordrecht
[141] Offerman T., Sonnemans J., Schram A. (1996) Economic Journal **106**:817–845
[142] van Dijk F., sonnemans J., van Winden F. (2002) Journal of Public Economics **85**:275–299
[143] Kuhlman D.M., Wimberley D.L. (1976) Journal of Personality and Social Psychology **34**:69–81
[144] Dawes R. (1989) Journal of Experimental Social Psychology **25**:1–17
[145] Dawes R., McTavish J., Shaklee H. (1977) Journal of Personality and Social Psychology **31(1)**:1–11
[146] Friedman D. (1991) Econometrica **59**:637–666
[147] Nowak M.A., Sigmund K. (1998) Nature **393**:573–577
[148] Axelrod R. (2000) Analyse & Kritik, special issue on the Evolution of Cooperation **22**:130–151
[149] Friedman M. (1953) in: *Milton Friedman, Essays in Positive Economics* The University of Chicago Press, Chicago 3–43
[150] Fehr E., Fischbacher U. (2003) Nature **425**:785–791
[151] Simon H.A. (1955) Quarterly Journal of Economics **69**:99–118
[152] Kahneman D., Knetsch J.L., Thaler R. (1986) American Economic Review **76**:728–741
[153] Trivers R. (1971) Quarterly Review of Biology **46**:35–57
[154] Boyd R., Richerson P.J. (1988) Journal of Theoretical Biology **132**:337–356
[155] Boyd R., Richerson P.J. (1992) Ethology and Sociobiology **13**:171–195
[156] Oliver P. (1980) American Journal of Sociology **85**:1356–1357
[157] Axelrod R. (1986) American Political Science Review **80**:1095–1111
[158] Yamagishi T., Hayashi N. (1994) in: U. Schulz, W. Albers, U. Mueller (eds.) *Social Dilemmas and Cooperation* Springer-Verlag, Berlin,Germany 311–326
[159] Henrich J., Boyd R. (2001) Journal of Theoretical Biology **208**:79–89
[160] Bowles S., Gintis H. (2003) The evolution of strong reciprocity: Cooperation in heterogeneous populations forthcoming in Theoretical Population Biology
[161] Cohen M.D., Riolo R.L., Axelrod R. (2001) Rationality and Society **13**:5–32

[162] Eshel I., Herreiner D., Samuelson L., Shaked E.S. (1998) Cooperation, mimesis and local interaction mimeo

[163] Burt R.S. (1999) Computational and Mathematical Organization Theory **5**:311–341

[164] Outkin A.V. (2003) Journal of Economic Behavior & Organization **52**:481–503

[165] Cohen M.D., Riolo R.L., Axelrod R. (1998) The emergence of social organization in the prisoner's dilemma: How context-preservationandother factors promote cooperation Santa Fe Institute Working Paper 99-01-002

[166] Stahl D.O., Wilson P.W. (1995) Games and Economic Behavior **10**:218–254

[167] Costa-Gomez M., Crawford V.P., Broseta B. (2001) Econometrica **69**:1193–1235

[168] Stahl D.O., Wilson P.W. (1994) Journal of economic behavior & organization **25**:309–327

[169] Rubinstein A. (1998) *Modeling Bounded Rationality* MIT Press, Cambridge, Massachusetts

[170] Jehiel P. (1995) Journal of Economic Theory **67**:497–519

[171] Smith A. (1975) *The Theory of Moral Sentiments* Clarendon Press, Oxford first published 1759

[172] Binmore K. (2000) *Game Theory and the Social Contract Volume I: Playing Fair* MIT Press, Cambridge, Massachusetts

[173] Ellison G. (1993) Econometrica **61(5)**:1047–1072

[174] Blume L.E. (1993) Games and economic behaviour **5**:387–424

[175] Fudenberg D., Maskin E. (1990) AEA Papers and Proceedings **80**:274–279

[176] Young H.P., Foster D. (1991) Games and Economic Behavior **3**:145–156

[177] Schuessler R. (1989) The Journal of Conflict Resolution **33**:728–749

[178] Vanberg V.J., Congleton R.D. (1992) American Political Science Review **86**:418–431

[179] Batali J., Kitcher P. (1995) Journal of Theoretical Biology **175**:161–171

[180] Hauert C., De Monte S., Hofbauer J., Sigmund K. (2002) Journal of Theoretical Biology **218**:187–194

[181] Morikawa T., Orbell J.M., Runde A.S. (1995) American Political Science Review **89**:601–611

[182] Orbell J., Runde A., Morikawa T. (1996) Biosystems **37**:81–97

[183] Mailath G.J., Samuelson L., Shaked A. (2001) in: U. Pagano, A. Nicita (eds.) *The Evolution of Economic Diversity* Routledge, New York 300–324
[184] Vega-Redondo F. (2002) Building up social capital in a changing world mimeo, Universidad de Alicante
[185] Flache A. (2001) Rationality and Society **13**:304–348
[186] Hayashi N., Yamagishi T. (1998) Personality and Social Psychology Review **2**:276–289
[187] Yamagishi T., Hayashi N., Jin N. (1994) in: U. Schulz, W. Albers, U. Mueller (eds.) *Social Dilemmas and Cooperation* Springer-Verlag, Berlin,Germany 233–250
[188] Ashlock D., Smucker M.D., Stanley E.A., Tesfatsion L. (1996) BioSystems **37**:99–125
[189] Young P.H. (1993) Econometrica **61**:57–84
[190] Kandori M., Mailath G.J., Rob R. (1993) Econometrica **61**:29–56
[191] Kandori M., Rob R. (1995) Journal Of Economic Theory **65**:383–414
[192] Samuelson L. (1994) Journal of economic theory **64**:35–65
[193] Foster D., Young P.H. (1990) Theoretical population biology **38**:219–232
[194] Young P.H. (1998) *Individual Strategy and Social Structure: An Evolutionary Theory of Institutions* Princeton University Press, New Jersey
[195] Davis D.D., Holt C.A. (1993) *Experimental Economics* Princeton University Press, Princeton, New Jersey
[196] Ostrom E., Walker J., Gardner R. (1992) American political science review **86**:404–417
[197] Yamagishi T. (1986) Journal of Personality and Social Psychology **51**:110–116
[198] Abreu D., Rubinstein A. (1988) Econometrica **56**:1259–1281
[199] Binmore K.G., Samuelson L. (1992) Journal of Economic Theory **57**:278–305
[200] Cooper D.J. (1996) Journal of Economic Theory **68**:266–275
[201] Cheung Y.W., Friedman D. (1997) Games and Economic Behavior **19**:46–76
[202] Kalai E., Lehrer E. (1993) Econometrica **61**:1019–1045
[203] Nachbar J.H. (1997) Econometrica **65**:275–309
[204] Fudenberg D., Levine D.K. (1998) *The Theory of Learning in Games* MIT Press, Cambridge, Massachusetts
[205] Roth A., Erev I. (1995) Games and economic behavior **8**:164–212
[206] Jehiel P. (1998) Games and economic behavior **22**:274–298

Index

Lecture Notes in Economics and Mathematical Systems

For information about Vols. 1–519 please contact your bookseller or Springer-Verlag

Vol. 520: B. Glaser, Efficiency versus Sustainability in Dynamic Decision Making. IX, 252 pages. 2002.

Vol. 521: R. Cowan, N. Jonard (Eds.), Heterogenous Agents, Interactions and Economic Performance. XIV, 339 pages. 2003.

Vol. 522: C. Neff, Corporate Finance, Innovation, and Strategic Competition. IX, 218 pages. 2003.

Vol. 523: W.-B. Zhang, A Theory of Interregional Dynamics. XI, 231 pages. 2003.

Vol. 524: M. Frölich, Programme Evaluation and Treatment Choise. VIII, 191 pages. 2003.

Vol. 525: S. Spinler, Capacity Reservation for Capital-Intensive Technologies. XVI, 139 pages. 2003.

Vol. 526: C. F. Daganzo, A Theory of Supply Chains. VIII, 123 pages. 2003.

Vol. 527: C. E. Metz, Information Dissemination in Currency Crises. XI, 231 pages. 2003.

Vol. 528: R. Stolletz, Performance Analysis and Optimization of Inbound Call Centers. X, 219 pages. 2003.

Vol. 529: W. Krabs, S. W. Pickl, Analysis, Controllability and Optimization of Time-Discrete Systems and Dynamical Games. XII, 187 pages. 2003.

Vol. 530: R. Wapler, Unemployment, Market Structure and Growth. XXVII, 207 pages. 2003.

Vol. 531: M. Gallegati, A. Kirman, M. Marsili (Eds.), The Complex Dynamics of Economic Interaction. XV, 402 pages, 2004.

Vol. 532: K. Marti, Y. Ermoliev, G. Pflug (Eds.), Dynamic Stochastic Optimization. VIII, 336 pages. 2004.

Vol. 533: G. Dudek, Collaborative Planning in Supply Chains. X, 234 pages. 2004.

Vol. 534: M. Runkel, Environmental and Resource Policy for Consumer Durables. X, 197 pages. 2004.

Vol. 535: X. Gandibleux, M. Sevaux, K. Sörensen, V. T'kindt (Eds.), Metaheuristics for Multiobjective Optimisation. IX, 249 pages. 2004.

Vol. 536: R. Brüggemann, Model Reduction Methods for Vector Autoregressive Processes. X, 218 pages. 2004.

Vol. 537: A. Esser, Pricing in (In)Complete Markets. XI, 122 pages, 2004.

Vol. 538: S. Kokot, The Econometrics of Sequential Trade Models. XI, 193 pages. 2004.

Vol. 539: N. Hautsch, Modelling Irregularly Spaced Financial Data. XII, 291 pages. 2004.

Vol. 540: H. Kraft, Optimal Portfolios with Stochastic Interest Rates and Defaultable Assets. X, 173 pages. 2004.

Vol. 541: G.-y. Chen, X. Huang, X. Yang, Vector Optimization. X, 306 pages. 2005.

Vol. 542: J. Lingens, Union Wage Bargaining and Economic Growth. XIII, 199 pages. 2004.

Vol. 543: C. Benkert, Default Risk in Bond and Credit Derivatives Markets. IX, 135 pages. 2004.

Vol. 544: B. Fleischmann, A. Klose, Distribution Logistics. X, 284 pages. 2004.

Vol. 545: R. Hafner, Stochastic Implied Volatility. XI, 229 pages. 2004.

Vol. 546: D. Quadt, Lot-Sizing and Scheduling for Flexible Flow Lines. XVIII, 227 pages. 2004.

Vol. 547: M. Wildi, Signal Extraction. XI, 279 pages. 2005.

Vol. 548: D. Kuhn, Generalized Bounds for Convex Multistage Stochastic Programs. XI, 190 pages. 2005.

Vol. 549: G. N. Krieg, Kanban-Controlled Manufacturing Systems. IX, 236 pages. 2005.

Vol. 550: T. Lux, S. Reitz, E. Samanidou, Nonlinear Dynamics and Heterogeneous Interacting Agents. XIII, 327 pages. 2005.

Vol. 551: J. Leskow, M. Puchet Anyul, L. F. Punzo, New Tools of Economic Dynamics. XIX, 392 pages. 2005.

Vol. 552: C. Suerie, Time Continuity in Discrete Time Models. XVIII, 229 pages. 2005.

Vol. 553: B. Mönch, Strategic Trading in Illiquid Markets. XIII, 116 pages. 2005.

Vol. 554: R. Foellmi, Consumption Structure and Macroeconomics. IX, 152 pages. 2005.

Vol. 555: J. Wenzelburger, Learning in Economic Systems with Expectations Feedback (planned) 2005.

Vol. 556: R. Branzei, D. Dimitrov, S. Tijs, Models in Cooperative Game Theory. VIII, 135 pages. 2005.

Vol. 557: S. Barbaro, Equity and Efficiency Considerations of Public Higer Education. XII, 128 pages. 2005.

Vol. 558: M. Faliva, M. G. Zoia, Topics in Dynamic Model Analysis. X, 144 pages. 2005.

Vol. 559: M. Schulmerich, Real Options Valuation. XVI, 357 pages. 2005.

Vol. 560: A. von Schemde, Index and Stability in Bimatrix Games. X, 151 pages. 2005.

Vol. 561: H. Bobzin, Principles of Network Economics. XX, 390 pages. 2006.

Vol. 562: T. Langenberg, Standardization and Expectations. IX, 132 pages. 2006.

Vol. 563: A. Seeger (Ed.), Recent Advances in Optimization. XI, 455 pages. 2006.

Vol. 564: P. Mathieu, B. Beaufils, O. Brandouy (Eds.), Artificial Economics. XIII, 237 pages. 2005.

Vol. 565: W. Lemke, Term Structure Modeling and Estimation in a State Space Framework. IX, 224 pages. 2006.

Vol. 566: M. Genser, A Structural Framework for the Pricing of Corporate Securities. XIX, 176 pages. 2006.

Vol. 567: A. Namatame, T. Kaizouji, Y. Aruga (Eds.), The Complex Networks of Economic Interactions. XI, 343 pages. 2006.

Vol. 568: M. Caliendo, Microeconometric Evaluation of Labour Market Policies. XVII, 258 pages. 2006.

Vol. 569: L. Neubecker, Strategic Competition in Oligopolies with Fluctuating Demand. IX, 233 pages. 2006.

Vol. 570: J. Woo, The Political Economy of Fiscal Policy. X, 169 pages. 2006.

Vol. 571: T. Herwig, Market-Conform Valuation of Options. VIII, 104 pages. 2006.

Vol. 572: M. F. Jäkel, Pensionomics. XII, 316 pages. 2006

Vol. 573: J. Emami Namini, International Trade and Multinational Activity. X, 159 pages, 2006.

Vol. 574: R. Kleber, Dynamic Inventory Management in Reverse Logistics. XII, 181 pages, 2006.

Vol. 575: R. Hellermann, Capacity Options for Revenue Management. XV, 199 pages, 2006.

Vol. 576: J. Zajac, Economics Dynamics, Information and Equilibnum. X, 284 pages, 2006.

Vol. 577: K. Rudolph, Bargaining Power Effects in Financial Contracting. XVIII, 330 pages, 2006.

Vol. 578: J. Kühn, Optimal Risk-Return Trade-Offs of Commercial Banks. IX, 149 pages, 2006.

Vol. 579: D. Sondermann, Introduction to Stochastic Calculus for Finance. X, 136 pages, 2006.

Vol. 580: S. Seifert, Posted Price Offers in Internet Auction Markets. IX, 186 pages, 2006.

Vol. 581: K. Marti; Y. Ermoliev; M. Makowsk; G. Pflug (Eds.), Coping with Uncertainty. XIII, 330 pages, 2006.

Vol. 582: J. Andritzky, Sovereign Default Risks Valuation: Implications of Debt Crises and Bond Restructurings. VIII, 251 pages, 2006.

Vol. 583: I.V. Konnov, D.T. Luc, A.M. Rubinov† (Eds.), Generalized Convexity and Related Topics. IX, 469 pages, 2006.

Vol. 584: C. Bruun, Adances in Artificial Economics: The Economy as a Complex Dynamic System. XVI, 296 pages, 2006.

Vol. 585: R. Pope, J. Leitner, U. Leopold-Wildburger, The Knowledge Ahead Approach to Risk. XVI, 218 pages, 2007 (planned).

Vol. 586: B.Lebreton, Strategic Closed-Loop Supply Chain Management. X, 150 pages, 2007 (planned).

Vol. 587: P. N. Baecker, Real Options and Intellectual Property: Capital Budgeting Under Imperfect Patent Protection. X, 276 pages , 2007.

Vol. 588: D. Grundel, R. Murphey, P. Panos , O. Prokopyev (Eds.), Cooperative Systems: Control and Optimization. IX, 401 pages , 2007.

Vol. 589: M. Schwind, Dynamic Pricing and Automated Resource Allocation for Information Services: Reinforcement Learning and Combinatorial Auctions. XII, 293 pages , 2007.

Vol. 590: S. H. Oda, Developments on Experimental Economics: New Approaches to Solving Real-World Problems. XVI, 262 pages, 2007.

Vol. 591: M. Lehmann-Waffenschmidt, Economic Evolution and Equilibrium: Bridging the Gap. VIII, 272 pages, 2007.

Vol. 592: A. C.-L. Chian, Complex Systems Approach to Economic Dynamics. X, 95 pages, 2007.

Vol. 593: J. Rubart, The Employment Effects of Technological Change: Heterogenous Labor, Wage Inequality and Unemployment. XII, 209 pages, 2007

Vol. 594: R. Hübner, Strategic Supply Chain Management in Process Industries: An Application to Specialty Chemicals Production Network Design. XII, 243 pages, 2007

Vol. 595: H. Gimpel, Preferences in Negotiations: The Attachment Effect. XIV, 268 pages, 2007

Vol. 596: M. Müller-Bungart, Revenue Management with Flexible Products: Models and Methods for the Broadcasting Industry. XXI, 297 pages, 2007

Vol. 597: C. Barz, Risk-Averse Capacity Control in Revenue Management. XIV, 163 pages, 2007

Vol. 598: A. Ule, Partner Choice and Cooperation in Networks: Theory and Experimental Evidence. Approx. 200 pages, 2007

Vol. 599: A. Consiglio, Artificial Markets Modeling: Methods and Applications. XV, 277 pages, 2007

Vol. 600: M. Hickman, P. Mirchandani, S. Voss (Eds.): Computer-Aided Scheduling of Public Transport. Approx. 424 pages, 2007

Vol. 601: D. Radulescu: CGE Models and Capital Income Tax Reforms: The Case of a Dual Income Tax for Germany. XVI, 168 pages, 2007

Vol. 602: N. Ehrentreich: Agent-Based Modeling: The Santa Fe Institute Artificial Stock Market Model Revisited. XVI, 225 pages, 2007

Vol. 603: D. Briskorn, Sports Leagues Scheduling: Models, Combinatorial Properties, and Optimization Algorithms. XII, 164 pages, 2008

Vol. 604: D. Brown, F. Kubler, Computational Aspects of General Equilibrium Theory: Refutable Theories of Value. XII, 202 pages, 2008

Vol. 605: M. Puhle, Bond Portfolio Optimization. XIV, 137 pages, 2008

Vol. 606: S. von Widekind, Evolution of Non-Expected Utility Preferences. X, 130 pages, 2008

Vol. 607: M. Bouziane, Pricing Interest Rate Derivatives: A Fourier-Transform Based Approach. XII, 191 pages, 2008

MIX
Papier aus verantwortungsvollen Quellen
Paper from responsible sources
FSC® C105338

If you have any concerns about our products,
you can contact us on
ProductSafety@springernature.com

In case Publisher is established outside the EU,
the EU authorized representative is:
Springer Nature Customer Service Center GmbH
Europaplatz 3, 69115 Heidelberg, Germany

Printed by Libri Plureos GmbH
in Hamburg, Germany